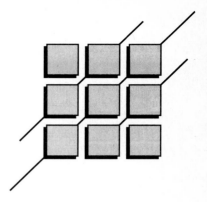

AUTOMATED MANUFACTURING SYSTEMS

ACTUATORS, CONTROLS, SENSORS, AND ROBOTICS

S. BRIAN MORRISS

CONESTOGA COLLEGE
ONTARIO, CANADA

McGraw-Hill

New York, New York   Columbus, Ohio   Mission Hills, California   Peoria, Illinois

Morriss, S. Brian.
    Automated Manufacturing Systems: Actuators, Controls, Sensors, and Robotics / S. Brian Morriss.
      p. cm.
    Includes index.
    ISBN 0-02-802331-5
    1. Automatic control.   2. Process control.   I. Title
TJ213.M553   1994
670.42'7—dc20   93-49739
                                                                                                                                        CIP

MS-DOS is a registered trademark and Windows NT is a trademark of Microsoft Corporation.
UNIX is a registered trademark of UNIX System Laboratories, Inc.
OS/2 is a registered trademark of International Business Machines Corporation.
Lotus is a registered trademark of Lotus Development Corporation.
WordPerfect is a registered trademark of WordPerfect Corporation.

Automated Manufacturing Systems: Actuators, Controls, Sensors, and Robotics

Copyright © 1995 by Glencoe/McGraw-Hill. All rights reserved. Except as permitted under the United States Copyright Act, no part of this publication may be reproduced or distributed in any form or by any means, or stored in a database or retrieval system, without prior written permission of the publisher.

Printed in the United States of America.

Send all inquiries to:
Glencoe/McGraw-Hill
936 Eastwind Drive
Westerville, OH 43081

ISBN 0-02-802331-5

1 2 3 4 5 6 7 8 9 10 11 12 13 14 15   RRD-C   99 98 97 96 95 94

CONTENTS

	PREFACE	x

PART A	**INTRODUCTION TO AUTOMATED SYSTEMS**		1

1 INTRODUCTION TO AUTOMATION — 2

Chapter Objectives — 2
1.0 AUTOMATE, EMIGRATE, LEGISLATE, OR EVAPORATE — 3
1.1 THE ENVIRONMENT FOR AUTOMATION — 3
    1.1.1 Automated Manufacturing, an Overview — 4
1.2 CONTROL OF AUTOMATION / PROCESS CONTROL — 6
1.3 COMPONENTS IN AUTOMATION — 8
    1.3.1 The Controller as an Automation Component — 8
    1.3.2 Sensors as Automation Components — 10
    1.3.3 Actuators as Automation Components — 11
1.4 INTERFACING AND SIGNAL CONDITIONING — 12
    1.4.1 Interfacing of Controllers to Controllers — 14
    1.4.2 Interfacing of Controllers to Other Components — 15
1.5 SUMMARY — 16
REVIEW QUESTIONS — 17

PART B	**AUTOMATION COMPONENTS**		19

2 ACTUATORS — 20

Chapter Objectives — 20
2.0 ACTUATORS — 21
2.1 LIGHT AND HEAT — 21
2.2 PIEZOELECTRIC FORCE GENERATORS — 25
2.3 SOLENOIDS AND TORQUE MOTORS — 25
2.4 AIR-POWER ACTUATORS AND SOLENOID-ACTUATED VALVES — 27
    2.4.1 Actuators — 27

	2.4.2 Valves	31
	2.4.3 Air Supply	33
2.5	HYDRAULIC ACTUATORS AND VALVES	35
2.6	CONVEYORS, FEEDERS, AND INDEXING TABLES	36
2.7	POWER TRANSMISSION COMPONENTS	39
2.8	SPECIAL-PURPOSE ACTUATOR SYSTEMS	41
2.9	SUMMARY	44
	REVIEW QUESTIONS	45

3 MOTORS 46

Chapter Objectives 46
3.1 THE HISTORY OF ELECTRIC MOTORS 47
3.2 CONSTRUCTION OF ELECTRIC MOTORS 48
3.3 THEORY OF OPERATION OF ELECTRIC MOTORS 50
3.4 TYPES OF ELECTRIC MOTORS 52
3.5 CONTROL OF MOTORS 53
    3.5.1 DC Motors for Control Applications 54
        Speed Control for DC Motors 55
        Stopping of DC Motors 57
        Methods of Controlling DC Power Levels 57
    3.5.2 AC Motors for Control Applications 58
        Control of AC Motors 65
        Stopping of AC Motors 68
    3.5.3 Electronically Commutated Motors 68
        Stepper Motors 69
        Timing Motors 69
        SCR Motors 70
        Brushless DC Servomotors 70
3.6 SUMMARY 73
REVIEW QUESTIONS 74

4 SENSORS 75

Chapter Objectives 75
4.0 SENSORS 76
4.1 QUALITY OF SENSORS 76

       4.1.1 Range and Span   76
       4.1.2 Error   76
             Resolution   77
             Repeatability   78
             Linearity   78
4.2 SWITCHES AND TRANSDUCERS   78
       4.2.1 Switches   79
       4.2.2 Non-Contact Presence Sensors (Proximity Sensors)   81
             PNP Versus NPN Sensors   82
       4.2.3 Temperature Transducers   86
       4.2.4 Force and Pressure Transducers   86
       4.2.5 Flow Transducers   88
4.3 POSITION SENSORS   89
4.4 VELOCITY AND ACCELERATION SENSORS   100
       4.4.1 Sensors Measuring Alternating Signals   102
4.5 SPECIAL-PURPOSE SENSOR SYSTEMS   104
4.6 SUMMARY   106
REVIEW QUESTIONS   107

5 VISION SYSTEMS   109

Chapter Objectives   109
5.0 VISION SYSTEMS   110
5.1 SIMPLE VISION SYSTEMS   110
5.2 THE COMPONENTS OF A VISION SYSTEM   113
       5.2.1 The Image   114
       5.2.2 The Camera   116
             The Camera Lens   116
             Image Conversion   117
       5.2.3 The Image Capturing System   120
             Sampling   121
             Quantization   122
       5.2.4 The Pre-Processor   123
             Grey Scale Reduction and Thresholding   124
             Windowing   127
             Image Coding   128

	5.2.5 The Memory	129
	5.2.6 The Processor	130
	Image Enhancement	130
	Object or Pattern Recognition	132
	Scene Analysis	140
	5.2.7 The Recognition Unit	142
	The Source of the Template	144
	5.2.8 The Output Interface	144
5.3	SUMMARY	145
REVIEW QUESTIONS		146

PART C   AUTOMATION CONTROLS   149

6 SERVOSYSTEMS   150

Chapter Objectives	150
6.0 SERVOSYSTEMS	151
6.1 CLOSED LOOP CONTROL SYSTEMS: A SIMPLE BLOCK DIAGRAM AND EXAMPLE EXPLAINED	151
6.2 SERVOSYSTEM ANALYSIS	155
6.2.1 Analysis by Transfer Function	156
6.2.2 Analysis by Graphical Methods	158
6.3 THE RESPONSE OF SERVOSYSTEMS	161
6.3.1 Stable Versus Unstable Response	161
6.3.2 Input Signals, Disturbance Types, and System Response	163
6.3.3 Physical Characteristics and System Response	166
6.3.4 Controller Algorithms and System Response	170
On/Off Control	170
Proportional Control (PE)	170
Proportional Integral Control (PI)	173
Proportional Derivative Control (PD)	175
Proportional Integral Derivative Control (PID)	178
Tuning of PID Control	180
6.4 THE CONTROLLER AND THE CONTROLLED SYSTEM	181
6.5 OTHER CONTROL CONFIGURATIONS	182
6.6 DIGITAL CONTROLLERS	186
6.7 SUMMARY	189
REVIEW QUESTIONS	190

7 CONTROLLERS: DIGITAL COMPUTERS AND ANALOG DEVICES — 192

- Chapter Objectives — 192
- 7.0 CONTROLLERS: DIGITAL COMPUTERS AND ANALOG DEVICES — 193
- 7.1 LEVELS OF CONTROL AND MAJOR COMPONENTS — 193
  - 7.1.1 Workcell Peripherals — 193
  - 7.1.2 Workcell Controller — 194
  - 7.1.3 Central Controller — 196
- 7.2 DIGITAL COMPUTERS, THE HEART OF AUTOMATED CONTROL — 197
  - 7.2.1 Computer Architecture — 197
  - 7.2.2 Computer Sizes — 199
  - 7.2.3 The CPU — 199
  - 7.2.4 Operating Systems, Software, and Memory Organization — 200
  - 7.2.5 Input/Output Devices — 204
    - Parallel I/O — 205
    - Analog I/O — 208
    - Serial I/O — 213
    - Multiplexing — 213
    - Communications and Protocol — 214
  - 7.2.6 Programmable Logic Controllers (PLCs) — 219
- 7.3 OTHER CONTROL COMPONENTS — 221
- 7.4 ANALOG CONTROLLERS — 225
  - 7.4.1 Analog Controllers Under Digital Control — 226
- 7.5 SUMMARY — 226
- REVIEW QUESTIONS — 227

PART D AUTOMATED SYSTEMS — 229

8 ROBOTS — 230

- Chapter Objectives — 230
- 8.0 ROBOTS — 231
- 8.1 ROBOTS AND NEAR-ROBOTS — 232
- 8.2 ROBOT GEOMETRIES — 233
  - 8.2.1 The Three Major Axes — 233
  - 8.2.2 Additional Degrees of Freedom — 235
- 8.3 ROBOT POWER SOURCES — 236
  - 8.3.1 Electric Motors — 236

8.3.2 Pneumatic Actuators	237
8.3.3 Hydraulic Actuators	238
8.4 COMPONENTS OF A COMPLETE ROBOT	239
8.4.1 The Arm	240
8.4.2 End Effectors	241
8.4.3 Peripherals	242
8.4.4 The Power Supply	243
8.4.5 The Controller	244
Central Processing Unit	244
Memory	245
Input/Output Capabilities	247
8.5 LEVELS OF SOPHISTICATION	249
8.5.1 Path Control	249
Point to Point Robots	250
Controlled Path Robots	251
Continuous Path Robots	251
8.5.2 Programming Languages	252
8.6 SUMMARY	253
REVIEW QUESTIONS	254

9 THE AUTOMATED WORKCELL AND ITS FUTURE — 255

Chapter Objectives	255
9.0 THE AUTOMATED WORKCELL AND ITS FUTURE	256
9.1 THE WORKCELL — AN EXAMPLE	257
What the Workcell Does...	257
What Each Controller Does...	258
9.2 ISLANDS OF AUTOMATION	264
9.3 FLEXIBLE MANUFACTURING SYSTEMS	266
9.3.1 The FMS Workcell — An Example	266
What the FMS Workcell Does...	266
9.4 COMPUTER INTEGRATED MANUFACTURING (CIM)	268
9.4.1 CIM — An Example	268
9.4.2 A Word from the Real World	271
9.5 SUMMARY	271
REVIEW QUESTIONS	272

10 AUTOMATION: WHEN? HOW? AND OTHER NON-TECHNICAL ISSUES — 273

Chapter Objectives	273
10.0 AUTOMATE, EMIGRATE, LEGISLATE, OR EVAPORATE	274
10.1 WHEN TO AUTOMATE	274
10.2 AN ORGANIZATIONAL APPROACH TO AUTOMATION	275
10.2.1 Simplification	276
JIT for Simplification	276
SPC for Simplification	278
10.2.2 Automation	278
10.2.3 Integration	278
10.3 SAFETY CONSIDERATIONS	279
10.4 COST JUSTIFICATION FOR AUTOMATION	280
10.5 SOCIAL EFFECTS OF AUTOMATION	284
The Problem	284
What Can We Do to Change Things?	286
10.5.1 The Problem	284
10.5.2 What Can We Do to Change Things?	286
10.6 SUMMARY	287
REVIEW QUESTIONS	288

PREFACE

This is a textbook about **automation.** It examines the components that can be used to build automated systems. **Actuators and sensors** are essential components, of course, and this text overviews sensors and actuators. This text also introduces the reader to computerized **controllers** as used in automated manufacturing. Some examples of complete automated systems are discussed.

Automation is still in its infancy. The future of automation belongs to the curious, to people who understand how components work and who can invent new ways of using simple principles. This text, therefore, concentrates on explaining the **principles of operation** of automation components and controllers. Some depth is sacrificed in the interest of providing a wide and balanced coverage of types of actuators, sensors, and controllers, and control theory. This information is supported by examples of applications in large and small automated systems.

The primary focus of this text is on the basics of automation. In order to place these basic concepts into context, this text also discusses some special-purpose automation components, automation systems, and automation concepts.

Bar code readers, numerical controlled manufacturing equipment, and *computer networks* are examples of the many special-purpose automation components that are discussed.

Computer aided design and *manufacturing software, database management,* and *computer integrated manufacturing* are some examples of the automation systems that this text covers.

Closed loop control, cost/quality justification for equipment, and automation implementation considerations are examples of the concepts important to automation that are discussed.

Three chapters in this text make it unusual as compared to other introductory-level automation textbooks. The chapter on **servosystems** is unique in its completeness and clarity. Today, when so many automated manufacturing processes require custom servo-control loops, the availability of this information to the average user is long overdue. The chapters on **motors** and **vision systems** are also unique in that they bring order to otherwise very complex and fragmented pockets of specialization.

Technical jargon has been avoided. While technical jargon can improve precision and efficiency, simple English is adequate for explaining principles. Simple English is better than technical precision if the target audience hasn't yet learned the technical "language."

This text does not require the reader to have a strong calculus background. Equations, where used, are as simple as possible. Even the chapter on servocontrol can be learned using the information in this text.

A course of study based on this text will provide readers with a solid *foundation*. After learning the material covered in this text, students will be able either to go directly into industry, able to communicate confidently with specialists and prepared for on-the-job learning of the final proprietary details for assembling, operating, and maintaining automated systems; or to go on to study specialized topics introduced in this text, having acquired a clear understanding of the operating principles of the components and systems peripheral to their chosen specialization.

Since this book examines the components and control of whole systems, and because it is written as an introductory text, it can be used in any of several ways. If this text is used in **an automated manufacturing technology program,** the instructor probably will not use the whole text. Many programs offer full courses on topics covered in single chapters here. For example, I don't use the Controller chapter because my students have had a full course on microprocessors by the time I see them. If they lacked training in microprocessors, I would have to use the Controller chapter. There is sufficient material in this text that whole chapters can be skipped and enough material will remain to easily fill a typical one semester course.

This text may be used as **the sole automation text.** For technicians, its content is adequate. Depending on the orientation of the technician program, the instructor may choose to teach the course in two semesters, or to bypass some of the more specialized chapters or sections (e.g., the Vision Systems chapter).

This text can be used **in a program where the course of study is already highly specialized in one aspect** of the material covered in this text. In an electronics technology program that emphasizes controls, for example, the chapters on actuators and sensors would be valuable, as would the chapter on servosystems.

This text also could be used **in a non-automation technologist or engineering program** to broaden the base of students. If, for example, a program is oriented toward mechanical product design, portions of this text can be used to ensure that students have the flexibility to diversify into equipment design. The instructor probably would bypass the chapters containing material covered elsewhere in the program.

The text is divided into four sections. Each section (in fact, each chapter) can be used alone. The four sections are:

Part A   Introduction to Automated Systems
A short examination of typical automated **workcells.** Component interaction in closed loop control is discussed. The **Computer Integrated Manufacturing** environment is introduced.

Part B   Automation Components
Chapters are dedicated to explaining the operation of simple **actuators** and of simple **sensors. Electric motors** are given their own chapter, where they are categorized and some techniques for driving them are discussed. **Vision systems, as complex sensors,** are covered in a separate chapter.

Part C   **Automation Control**
Digital computers, including **programmable controllers,** are described. Interfacing of controllers to actuators and to sensors is discussed. **Analog controllers and components** are covered. A chapter on **servocontrol** describes applications and response characteristics of servo-controlled systems.

Part D   **Automation Systems**
Controlled systems are discussed here. This section begins by examining an "off-the-shelf" automated system, the **industrial robot.** Another chapter examines examples of a typical **workcell,** a **Flexible Manufacturing System,** and a **Computer Integrated Manufacturing system** in more detail than in Part A. A final chapter explores issues peripheral to automation. These issues include **economic justification** for automated systems, techniques to involve existing workforces in the **implementation of automation,** and a discussion on the **impact of automation** on individuals and on organizations.

ACKNOWLEDGMENTS

There are several people whom I would like to acknowledge. I would like to thank the students of Conestoga College in the Automated Manufacturing Mechanical Technology program, in the Robotics and Automation Mechanical Engineering Technology program, and in the Electronics Technology—Computer Option program, who had to use the early drafts of this text in their course work. They never failed to help me find sections that needed revision.

Thanks go to the administration of Conestoga College, who made it possible for me to participate in the development of whole automation-related programs. This participation encouraged me to develop a broad view of the material a student would need to study, and made it possible/necessary for me to investigate areas where my industrial experiences hadn't taken me, or where there were new developments.

I would also like to thank my friends and contacts in industry. Jerry Wootten, a systems integrator with Automation Tooling Systems, has shared his views regarding trends in automation with me in hundreds of discussions. John Tielmans, whose professional career has always revolved around process control, reviewed the servosystems chapter for me, and participated in many more technical discussions. Numerous other friends and contacts have discussed their field of expertise, allowing me to expand my horizons and keep on top of new developments.

Several contacts in industry have donated automation equipment and software to Conestoga College, equipment that has affected this text. I would like to mention Alex Scott, of Syspro Impact Software Inc, who supplied the college with MRP II software; Don Gordon, of Westinghouse Electric Corporation, who supplied state-of-the-art PLCs and I/O modules; Jon Olley, of Budd Automotive, who donated a whole truckload of miscellaneous sensors, actuators, and controllers; and many other donors who are too numerous to mention.

Of several colleagues who have used the developing manuscript to teach courses of my design, and who have provided me with valuable feedback, Taylor Morey provided the most exhaustive critique. Pat Tondreau reviewed the final chapter, particularly the section on cost justification, and provided valuable feedback. The author is grateful to Ray Beets, Iowa Central Community College, Fort Dodge, Iowa; Larry Catron, Scott County Vocational Center; Gate City, Virginia; Harry Presley, Itawamba Community College, Tupelo, Mississippi; and Dr. Lee Rosenthal, Fairleigh Dickenson University, Teaneck, New Jersey, for their recommendations for topical coverage and their detailed, technical, and practical suggestions.

Finally, I would like to thank my wife, who hates anything related to technology yet who understood (sometimes) why I had to spend so many hours writing at "that stupid computer," or working late with "those stupid robots." She helped me to remember that those who don't care about technology might still be wonderful people. I'm finished now, dear. I can help you set the clock on the VCR.

S. Brian Morriss

Graphic marks have been inserted into the text. At the ☑ the reader should pause to consider what has just been read before going on to the next idea in this text.

PART A

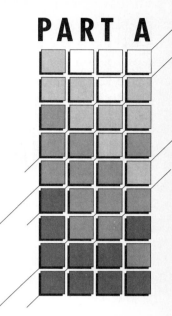

INTRODUCTION TO AUTOMATED SYSTEMS

1	INTRODUCTION TO AUTOMATION	2

CHAPTER 1

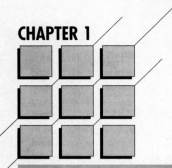

INTRODUCTION TO AUTOMATION

CHAPTER OBJECTIVES

After completing this chapter, the reader will be able to:

- Discuss the challenges facing an organization contemplating an automation program.
- Describe the three steps in moving toward a Computer Integrated Manufacturing environment.
- Describe the components in a Computer Integrated Manufacturing environment.
- Discuss the difference between:
  - open versus closed loop control
  - automation control versus process control
  - hard versus soft automation
- Identify the essential components of a closed-loop controlled system, and describe the operation of simple closed-loop controlled systems.
- Describe the benefits of digital computers, including Programmable Controllers, as automation and process controllers.
- Discuss the interaction of sensors, controllers, and actuators and the reasons for signal conditioning.
- Describe the development of technology which enables automated manufacturing, and discuss the current limitations of the technology.

1.0 AUTOMATE, EMIGRATE, LEGISLATE, OR EVAPORATE

"Automate, emigrate, legislate or evaporate." This was a choice many North American manufacturers faced in the 1980s. Offshore competitors were producing goods more cheaply, and with higher quality, than were North American manufacturers. Customers were buying Toyotas instead of Fords, Sonys instead of RCA Victors.

Some manufacturers tried to lower prices by reducing manufacturing costs. They either automated or emigrated. Many countries legislated trade barriers to keep high quality, low cost products out. Manufacturers who did nothing … disappeared, often despite their own government's protective trade barriers.

Many consumers still choose imports over domestic products, but some North American manufacturers are now trying more thoughtful measures to meet the challenge.

Automation is a technique that can be used to reduce costs and/or to improve quality. Automation can increase manufacturing speed, while reducing cost. Automation can lead to products having consistent quality, perhaps even consistently good quality.

Some manufacturers who automated survived. Others didn't. The ones who survived were those who used automation to *improve quality*. It often happened that improving quality led to reduced costs.

1.1 THE ENVIRONMENT FOR AUTOMATION

Automation, the subject of this textbook, is not a magic solution to financial problems. It is, however, a valuable tool that can be used to improve product quality. Improving product quality, in turn, results in lower costs. Producing inexpensive, high quality products is a good policy for any company.

But where do you start?

Simply considering an automation program can force an organization to face problems it might not otherwise face:

- What automation and control technology is available?
- Are employees ready and willing to use new technology?
- What technology should we use?
- Should the current manufacturing process be improved before automation?
- Should the product be improved before spending millions of dollars acquiring equipment to build it?

Automating before answering the above questions would be foolish. The following chapters describe the available technology so that the reader will be prepared to select appropriate automation technology. The answers to the last two questions above are usually "yes," and this book introduces techniques to improve processes and products, but each individual organization must find its own improvements.

1.1.1 Automated Manufacturing, an Overview

Automating of individual manufacturing cells should be the second step in a three step evolution to a different manufacturing environment. These steps are:

1. **Simplification** of the manufacturing process. If this step is properly managed, the other two steps might not even be necessary. The "Just In Time" (JIT) manufacturing concept includes procedures that lead to a simplified manufacturing process. See chapter 10 for more details.

2. **Automation** of individual processes. This step, the primary subject of this text, leads to the existence of "islands of automation" on the plant floor. The learning that an organization does at this step is valuable. An organization embarking on an automation program should be prepared to accept some mistakes in the early stage of this phase. The cost of those mistakes is the cost of training employees.

3. **Integration** of the islands of automation and other computerized processes into a total manufacturing and business system. While this text does not discuss the details of integrated manufacturing, it is discussed in general in this chapter and again in chapter 10. Technical specialists should be aware of the potential future need to integrate, even while they embark on that first "simplification" step.

The large, completely automated and integrated environment shown in figure 1.1 is a Computer Integrated Manufacturing (CIM) operation. The CIM operation includes:

- **Computers,** including:
  - one or more "host" computers
  - several cell controller computers
  - a variety of personal computers
  - Programmable Controllers (PLC)
  - computer controllers built into other equipment

- **Manufacturing Equipment,** including:
  - robots
  - numerical control machining equipment (NC, CNC, or DNC)
  - controlled continuous process equipment (e.g., for turning wood pulp into paper)
  - assorted individual actuators under computer control (e.g., motorized conveyor systems)
  - assorted individual computer-monitored sensors (e.g., conveyor speed sensors)
  - pre-existing "hard" automation equipment, not properly computer-controllable, but monitored by retro-fitted sensors

- **Computer Peripherals,** such as:
  - printers, FAX machines, terminals, paper-tape printers, etc.

- **Local Area Networks (LANs)** interconnecting computers, manufacturing equipment, and shared peripherals. "Gateways" may interconnect incompatible LANs and incompatible computers.

- **Databases (DB),** stored in the memories of several computers, and accessed through a **Database Management System (DBMS)**

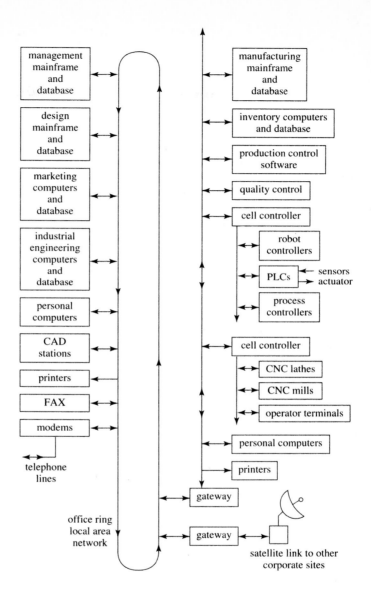

Fig. 1.1 A computer integrated manufacturing environment

- **Software Packages** essential to the running of the computer system. Essential software would include: the operating system(s) (e.g., UNIX, WINDOWS NT, OS/2 or SNA) at each computer; communications software (e.g., the programs which allow the LANs to move data from computer to computer and which allow some controllers to control other controllers); and the database management system (DBMS) programs.

- **assorted "processes"** which use the computers. These computer- users might include:
  - humans, at computer terminals, typing commands or receiving information from the computers
  - Computer Aided Design (CAD) programs
  - Computer Aided Engineering (CAE) programs
  - Computer Aided Manufacturing (CAM) programs, including those that do production scheduling (MRP), process planning (CAPP), and monitoring of shop floor feedback and control of manufacturing processes (SFC)
  - miscellaneous other programs for uses such as word processing or financial accounting

As of this writing, very few completely computer integrated manufacturing systems are in use. There are lots of partially integrated manufacturing systems. Before building large computer integrated systems, we must first understand the components and what each component contributes to the control of a simple process.

1.2 CONTROL OF AUTOMATION/PROCESS CONTROL

Automated processes can be controlled by humans operators, by computers, or by a combination of the two. If a human operator is available to monitor and control a manufacturing process, **open loop control** may be acceptable. If a manufacturing process is automated, then it requires **closed loop control.**

Figure 1.2 shows examples of open loop control and closed loop control. One major difference is the presence of the sensor in the closed loop control system. The motor speed controller uses the feedback it receives from this sensor to verify that the speed is correct, and drives the actuator harder or softer until the correct speed is achieved. In the open loop control system, the operator uses his/her built-in sensors (eyes, ears, etc.) and adjusts the actuator (via dials, switches, etc.) until the output is correct. Since the operator provides the sensors and the intelligent control functions, these elements do not need to be built into an open loop manufacturing system.

Human operators are more inconsistent than properly programmed computers. Anybody who has ever shared the road with other drivers is familiar with the disadvantages of human control. Computerized controls, however, can also make mistakes, when programmed to do so. Programming a computer to control a complex process is very difficult (which is why human automobile operators have not yet been replaced by computers).

The recent development of affordable digital computers has made automation control possible. Process control has been around a little longer. The difference in the meanings of these two terms is rapidly disappearing.

Process control usually implies that the product is produced in a continuous stream. Often, it is a liquid that is being processed. Early process control systems consisted of specially-designed analog control circuitry that measured a system's output (e.g., the temperature of liquid leaving a tank), and changed that output (e.g., changing the amount of cool liquid mixed in) to force the output to stay at a preset value.

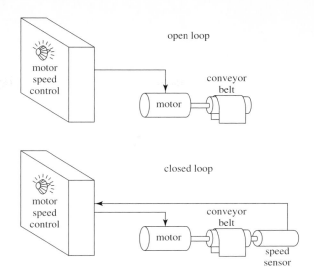

Fig. 1.2 Open loop and closed loop speed control

Automation control usually implies a sequence of mechanical steps. A camshaft is an automation controller because it mechanically sequences the steps in the operation of an internal combustion engine. Manufacturing processes are often sequenced by special digital computers, known as programmable logic controllers (PLCs), which can detect and can switch electrical signals on and off. Digital computers are ideally suited for "automation control" type tasks, because they consist of circuits each of which can only be either "on" or "off."

Process control is now usually accomplished using digital computers. Digital controllers may be built into cases with dials and displays which make them look like their analog ancestors. PLCs can also be programmed to operate as analog process controllers. They offer features which allow them to measure and change analog values. Robots and NC equipment use digital computers and a mixture of analog and digital circuit components to control "continuous" variables such as position and speed.

Advances in automation and process control have been rapid since the start of the silicon revolution. Before modern silicon devices, controllers were built for specific purposes and could not be altered easily. A camshaft sequencer would have to have its camshaft replaced to change its control "program." Early analog process controllers had to be rewired to be reprogrammed. Automation systems of these types are called **hard automation.** They do what they are designed and built to do, quickly and precisely perhaps, but with little adaptability for change (beyond minor adjustments). Modification of hard automation is time-consuming and expensive, since modifications can only be performed while the equipment sits idle.

As digital computers and software improve, they are replacing hard automation. Digital computer control gives us **soft automation.** Modern digital computers are reprogrammable. It is even possible to reprogram them and test the changes while they work!

Even if hardware changes are required to a soft automation system, the lost time during changeover is less than for hard automation. Even a soft automation system has to be stopped to be retro-fitted with additional sensors or actuators, but the controller doesn't have to be rebuilt to use these added pieces.

Digital computers are cheap, powerful, fast and compact. They offer several new advantages to the automation user. A single digital controller can control several manufacturing processes. The designer only needs to ensure the computer can monitor and control all processes quickly enough, and has some excess capacity for future changes.

Using digital computers, the equipment that is to be controlled can be built to be more "flexible." A computer controlled milling machine, for example, can come equipped with several milling cutters and a device to change them. The computer controller can include programs that exchange cutters between machining operations.

Soft automation systems can be programmed to detect and to adapt to changes in the work environment or to changes in demand. An NC lathe can, for example, modify its own cutting speed if it detects a sudden change in the hardness of a raw material being cut. It may also change its own programming in response to a signal from another automated machine requesting a modification in a machined dimension.

1.3 COMPONENTS IN AUTOMATION

When we discuss automation in this text, we will always mean controlled automation. A circuit that allows you to control a heater is a labor-saving device, but without components to ensure that the room temperature remains comfortable, it is not an automated system.

Figure 1.3 illustrates the essential components in a controlled automated system:

- the **actuator** (which does the work)
- the **controller** (which "tells" the actuator to do work)
- and the **sensor** (which provides feedback to the controller so that it knows the actuator is doing work)

1.3.1 The Controller as an Automation Component

A controlled system may be a simple digital system. An example is shown in figure 1.4, in which the actuator consists of a pneumatic valve and a pneumatic cylinder that must be

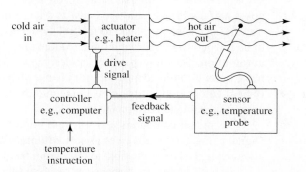

Fig. 1.3 Components of a simple controlled automation system

either fully extended or retracted. The controller is a PLC that has been programmed to extend the cylinder during some more complicated process, and to go on to the next step in the process *only* after the cylinder extends.

When it is time to extend the cylinder, the PLC supplies voltage to the valve, which *should* open to provide air to the cylinder, which *should* then extend. If all goes well, after a short time the PLC will receive a change in voltage level from the limit switch, allowing it to execute the next step in the process. If the voltage from the switch does not change for any reason (faulty valve or cylinder or switch, break in a wire, obstruction preventing full cylinder extension, etc.), the PLC will not execute the next step. The PLC may even be programmed to turn on a "fault" light when such a delay occurs.

A controlled system might be an analog system, as illustrated in figure 1.5. In this system, the actuator is an hydraulic servovalve, and a fluid motor. The servovalve opens proportionally with the voltage it receives from the controller, and the fluid motor rotates faster if it receives more hydraulic fluid. There is a speed sensor connected to the motor shaft, which outputs a voltage signal proportional to the shaft speed. The controller is programmed to move the output shaft at a given speed until a load is at a given position.

When the program requires the move to take place, the controller outputs an *approximately* correct voltage to the servovalve, then monitors the sensor's feedback signal. If the speed sensor's output is different than expected (indicating wrong motor speed), the controller increases or decreases the voltage supplied to the servovalve until the correct feedback voltage is achieved. The motor speed is controlled until the move finishes. As with the digital control example, the program may include a function to notify a human operator if speed control isn't working.

Digital and analog controllers are available "off the shelf" so that systems can be constructed inexpensively (depending on your definition of "inexpensive"), and with little specialized knowledge required.

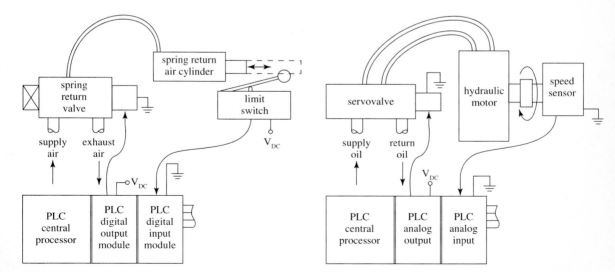

Fig. 1.4 A digital controlled system   **Fig. 1.5** An analog controlled system

Most controllers include **communication ports** so that they can send or receive signals and instructions from other computers. This allows individual controllers to be used as parts of **distributed control** systems, in which several controllers are interconnected. In distributed control, individual controllers are often slaves of other controllers, and may control slave controllers of their own.

1.3.2 Sensors as Automation Components

Obviously, controlled automation requires devices to sense system output. Sensors also can be used so that a controller can detect (and respond to) changing conditions in its working *environment*. Figure 1.6 shows some conditions that a fluid flow control system might have to monitor. Sensors are required to sense three settings for values that have to be controlled (flow rate, system pressure, and tank level) and to measure the actual values of those three variables. Another sensor measures the uncontrolled input pressure, so that the valve opening can be adjusted to compensate.

A wide range of sensors exists. Some sensors, known as **switches,** detect when a measured condition exceeds a pre-set level (e.g., closes when a workpiece is close enough to work on). Other sensors, called **transducers,** can describe a measured condition (e.g., output increased voltage as a workpiece approaches the working zone).

Sensors exist that can be used to measure such variables as:

- presence or nearness of an object
- speed, acceleration, or rate of flow of an object
- force or pressure acting on an object
- temperature of an object
- size, shape, or mass of an object
- optical properties of an object
- electrical or magnetic properties of an object

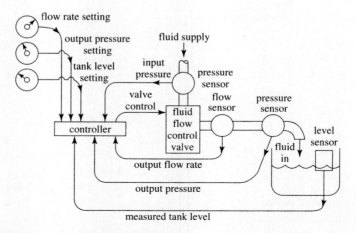

Fig. 1.6 Sensing in an automated system

The operating principles of several sensors will be examined in Chapter 4.

Sensors may be selected for the *types of output* they supply. Some sensors output DC voltage. Other sensors output current proportional to the measured condition. Still other sensors have AC output. Some sensors, which we will not examine in this text, provide non-electrical outputs. Since a sensor's output may not be appropriate for the controller that receives it, the signal may have to be altered or "conditioned." Some sensor suppliers also sell **signal conditioning** units. Signal conditioning will be discussed later in this chapter, and in more detail in Chapter 7.

1.3.3 Actuators as Automation Components

An actuator is controlled by the "controller." The actuator, in turn, changes the output of the automated process. The "actuator" in an automated process *may in fact be several actuators,* each of which provides an output that drives another in the series of actuators. An example can be found in figure 1.7, in which an hydraulic actuator controls the position of a load.

The controller outputs a low current DC signal. The signal goes to the first stage of actuator, the amplifier, which outputs increased voltage and current. The large DC power supply and the amplifier are considered part of the actuator. The amplified DC causes the second stage of the actuator, the hydraulic servovalve, to open and allow hydraulic fluid flow, proportional to the DC it received. The servovalve, hydraulic fluid, pump, filter, receiving tank and supply lines are all components in the actuator. The fluid output from the valve drives the third actuator stage, the hydraulic cylinder, which moves the load.

A servovalve is called a *servo*valve because it contains a complete closed-loop position control system. The internal control system ensures that the valve opens proportionally with the DC signal it receives. The three-stage actuator we have been discussing as a component of a closed loop control system, therefore, includes another closed-loop control system.

 Some actuators can only be turned **on** or **off.** A heater fan helps to control temperature when it is turned on or off by a temperature control system. Pneumatic cylinders are usually either fully extended ("on") or fully retracted ("off"). Other actuators respond **proportionally** with the signal they receive from a controller. A variable speed motor is

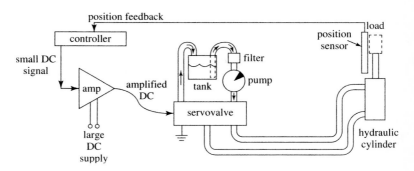

Fig. 1.7 Hydraulic actuated position control

an actuator of this type. Hydraulic cylinders can be controlled so that they move to positions between fully extended and fully retracted.

Actuators can be purchased to change such variables as:
- presence or nearness of an object
- speed, acceleration, or rate of flow of an object
- force or pressure acting on an object
- temperature of an object
- final machined dimensions of an object

The operating principles of several actuators will be examined in Chapter 2.

 Actuators can be selected for the types of *inputs* they require. Some actuators respond to DC voltage or current. Other actuators require AC power to operate. Some actuators have two sets of input contacts: one set for connecting a power supply, and the other set for a low power signal that tells the actuator how much of the large power supply to use. An hydraulic servovalve is an actuator of this type. It may be connected to a 5000 PSI hydraulic fluid supply, capable of delivering fluid at 10 gallons per minute, and also to a 0 to 24 volt DC signal which controls how much fluid the valve actually allows through.

As with sensors, there may be incompatibilities between the signal requirement of an actuator and a controller's inherent output signals. Signal conditioning circuitry may be required.

1.4 INTERFACING AND SIGNAL CONDITIONING

There was quite a long delay before digital computer control of manufacturing processes became widely implemented. Lack of standardization was the problem.

NC equipment showed up as the first real application of digital control. The suppliers of NC equipment built totally enclosed systems, evolving away from analog control toward digital control. With little need for interconnection of the NC equipment to other computer controlled devices, the suppliers did not have to worry about lack of standards in communication. Early NC equipment read punched tape programs as they ran. Even the paper tape punchers were supplied by the NC equipment supplier.

Meanwhile, large computer manufacturers, such as IBM and DEC, concentrated on interconnecting their own proprietary equipment to their own proprietary office peripherals. (Interconnection capability was poor even there.)

Three advances eventually opened the door for automated manufacturing by allowing easier interfacing of controllers, sensors and actuators. One advance was the *development of the programmable controller,* then called a "PC," now called a "PLC" (programmable logic controller). PLCs contain digital computers. It was a major step from sequencing automation with rotating cams or with series of electrical relay switches, to using microprocessor-based PLC sequencers. With microprocessors, the sequencers could be programmed to follow different sequences under different conditions.

The physical structure of a PLC, as shown in figure 1.8, is as important a feature as its computerized innards. The central component, called the *CPU,* contains the digital computer and plugs into a *bus* or a *rack.* Other PLC modules can be plugged into the same bus. Optional *interface modules* are available for just about any type of sensor or actuator. The PLC user buys only the modules needed, and thus avoids having to worry about compatibility between sensors, actuators and the PLC. Most PLCs offer communication modules now, so that the PLC can exchange data with at least other PLCs of the same make. Figure 1.9 shows a PLC as it might be connected for a position-control application: reading digital input sensors, controlling AC motors, and exchanging information with the operator.

Another advance which made automation possible was the *development of the robot.* A variation on NC equipment, the robot in figure 1.10 is a self-enclosed system of actuators, sensors and controller. Compatibility of robot components is the robot manufacturer's problem. A robot includes built-in programs allowing the user to "teach" positions to the arm, and to play-back moves. Robot programming languages are similar to other computer programming languages, like BASIC. Even the early robots allowed connection of certain types of external sensors and actuators, so complete work cells could be built around, and controlled by, the robot. Modern robots usually include communication ports, so that robots can exchange information with other computerized equipment with similar communication ports.

The third advance was *the introduction, by IBM, of the personal computer (PC).* (IBM's use of the name "PC" forced the suppliers of programmable controllers to start calling their "PC"s by another name, hence the "PLC.") IBM's PC included a feature then called **open architecture.** What this meant was that inside the computer box was a computer on a single "mother" circuit-board and several slots on this "motherboard." Each slot is a standard connector into a standard bus (set of conductors controlled by the computer). This architecture was called "open" because IBM made information available so that other manufacturers could design circuit boards that could be plugged into the bus. IBM also provided information on the operation of the motherboard so that others

(a) (b)

Fig. 1.8 (a) A programmable controller; (b) installation of I/O module on bus unit. (Photographs by permission, Westinghouse Electric Corporation, Electrical Components Division, Pittsburgh, Pennsylvania.)

could write IBM PC programs to use the new circuit boards. IBM undertook to avoid changes to the motherboard that would obsolete the important bus and motherboard standards. Boards could be designed and software written to interface the IBM PC to sensors, actuators, NC equipment, PLCs, robots, or other computers, without IBM having to do the design or programming. Coupled with IBM's perceived dependability, this "open architecture" provided the standard that was missing. Now computerization of the factory floor could proceed.

1.4.1 Interfacing of Controllers to Controllers

Standards for what form the signals should take for communication between computers are still largely missing. The PLC, robot, and computer manufacturers have each developed their own standards, but one supplier's equipment can't communicate very easily with another's.

Most suppliers do, at least, build their standards around a very basic set of standards which dates back several decades to the days of teletype machines: the **RS 232 standard.** Because of the acceptance of RS 232, a determined user can usually write controller programs which exchange simple messages with each other.

The International Standards Organization (ISO) is working to develop a common communication standard, known as the OSI (Open Systems Interconnectivity) model. Several commercial computer networks are already available, many using the agreed-on

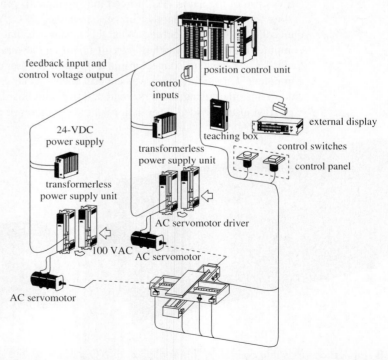

Fig. 1.9 A PLC in a position control application. (Illustration by permission, OMRON Canada Inc., Scarborough, Ontario, Canada.)

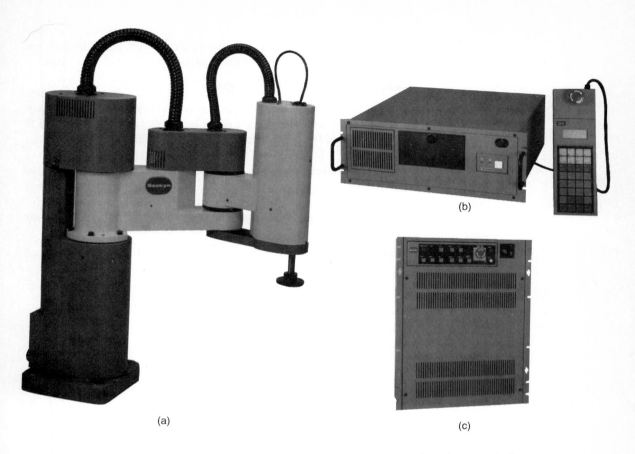

Fig. 1.10 (a) An industrial robot; (b) robot controller; and (c) power supply. (Photographs by permission, Sankyo Seiki (America) Inc., Robotics Division, Boca Raton, Florida.)

parts of the OSI model. Manufacturer's Automation Protocol (MAP) and Technical and Office Protocol (TOP) are the best-known of these.

Despite the recognition that common standards are needed to be recognized, the immediate need for communications is leading to the growth of immediately available **proprietary standards.** Several large PLC manufacturers sell proprietary local area networks and actively encourage others to join them in using those standards. Simultaneous growth of proprietary office local area network suppliers means that interconnecting the plant to the manufacturing office is becoming another problem area. One promising aspect of the growth of giants in the local area network field is that the giants recognize the need for easy-to-use interfaces between their systems, and have the money to develop them.

1.4.2 Interfacing of Controllers to Other Components

One area in which development of standards is not as great a problem is the connection of sensors and actuators to controllers. This area is a less severe problem because electrical actuators have been available for so long that several standards are effectively in force.

The 24 volt DC solenoid-actuated valve is an example. 240 volt three phase AC is another standard, dictated by power supply utility companies. Sensor suppliers, more recent arrivals, have simply adopted those standards most appropriate to their target market.

Lack of a *single* standard is still an inconvenience. Signal conditioning is often required so that incompatible components and controllers can be interconnected. The size of the problem can be reduced by the user by selecting components with similar power requirements and control signal characteristics, if possible. Another option is using a PLC as the controller, and selecting a PLC which offers I/O modules for all the different sensors and actuators to be used.

The user could also consider employing popular "open architecture" computers that can be retrofitted with interfacing circuit cards available from several sources; however, programming the control program to use those interface cards requires some skill.

Another alternative is buying or building signal conditioning interface circuits to do signal modification such as:

- cleaning noisy electrical signals
- isolating high power signals from low power signals
- amplification (or de-amplification) of current or voltage levels
- converting analog signals to/from digital numbers
- converting DC signals to/from AC signals
- converting electrical signals to/from non-electrical signals

SUMMARY

Automation can be used to reduce manufacturing **costs.** It is more appropriately used to improve **quality** and make it more consistent.

This chapter looked at a **Computer Integrated Manufacturing** environment, so that users of this book will be able to anticipate the potential third step in the **simplify-automate-integrate** process as they design individual islands of automated manufacturing.

An automated process must be able to **measure and control** its output. "Closed loop control" is the term used for a self-controlled automated process. A complete closed loop control system requires an **actuator** (perhaps several working together), which responds to signals from a **controller,** and at least one **sensor** that the controller uses to ensure that the manufacturing process is proceeding as it should. Most controlled automation processes use several sensors, to detect incoming materials and workplace environmental conditions, as well as to measure and verify the process's output.

Digital computers are used in modern **soft automation** systems, replacing hard automation controllers. Custom-built analog controllers, traditionally used for process control applications, are examples of "hard" automation. The main advantages to be reaped by using digital control are in the area of increased **flexibility.** Digital controllers, including PLCs, can be programmed to respond to more conditions, in more ways, and can be reprogrammed more easily than hard automation. The initial system design and building steps are easier, too, because off-the-shelf components can be used.

Numerical controlled (NC) machining components, analog process controllers, and mechanical and electrical sequencers have been available for quite a while, but the development of **programmable controllers** (PLCs), **robots**, and **open-architecture computers** really made automation accessible to the average industrial user. **Signal conditioning** is still required to get some components to work with other components. While computerized equipment can be programmed to communicate with each other, **standardization in communication** between computerized equipment is not fully realized yet.

REVIEW QUESTIONS

1. What questions might an organization be faced with as it first contemplates automating?
2. What three steps (list in order) should an organization be prepared to take when it decides to embark on a program of automation?
3. List the types of components that make up a completely computer integrated manufacturing environment.
4. How is closed loop control different from open loop control? Where is closed loop control preferable?
5. What is the implied difference between "automation control" and "process control?" Is the difference still a real difference?
6. What is the difference between hard automation and soft automation? Which would be preferable in a plant producing products for a rapidly changing marketplace? Why?
7. What are the three essential components in a controlled automated system? Briefly state what each component contributes.
8. A switch is one type of sensor. What is the other type and how does it differ from a switch?
9. Give one example of a manufacturing situation where you might use an actuator that is simply switched on or off, and one example of where an actuator that can be controlled between fully on or fully off would be a better choice.
10. What is meant by "open architecture" as the term is used to describe a computer?
11. How would deciding to use a programmable controller (PLC) make the design of an automated system easier?
12. The ISO has a subgroup known as the OSI subgroup. What do these letters stand for? What is the OSI subgroup doing for automation, and why is its work important?

PART B

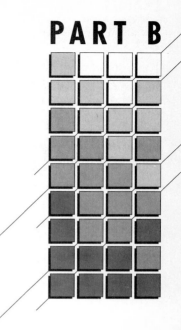

AUTOMATION COMPONENTS

2	ACTUATORS	20
3	MOTORS	46
4	SENSORS	75
5	VISION SYSTEMS	109

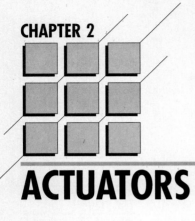

CHAPTER 2

ACTUATORS

CHAPTER OBJECTIVES

After completing this chapter, the reader will be able to:

- Describe the principles of operation of a wide range of actuators, including low power types such as:
  - resistance heaters
  - Light Emitting Diodes (LED)
  - Liquid Crystal Displays (LCD)
  - gas plasma displays
  - Cathode Ray Tubes (CRT)
  - piezoelectric force generators

  Medium power types such as:
  - solenoids
  - torque motors
  - air cylinders and hydraulic cylinders
  - fluid-powered rotary actuators
  - vacuum generators
  - pneumatic valves
  - hydraulic valves

- Select from a wide range of automation components and power moving devices, including:
  - chains for automation conveyors
  - ballscrews and ballnuts
  - vibratory feeders
  - indexing mechanisms
  - harmonic gearing

■ Describe special purpose automated systems that can be used as actuators, including:
- X-Y tables
- NC machinery
- circuit card component inserters, soldering machines, and testers
- Automated Storage and Retrieval Systems (AS/RS)

2.0 ACTUATORS

An actuator performs work. Controlled by a computer, it provides the output of an automated system.

There is a wide range of actuators available for designers to choose from. In this chapter, we will examine the characteristics of some, but definitely not all, types of actuators. We will first examine the simplest and most common, then move on to cover some of the more unique and special-purpose actuators. A later chapter (chapter 3) will cover motors. Still other chapters will discuss integrating actuators with other components to build automated workcells, flexible manufacturing systems (FMS), and even fully computer integrated manufacturing (CIM) systems.

2.1 LIGHT AND HEAT

Electric heaters have been used for years. If you pass current through a resistor, heat is generated. Resistance wire is coiled into tubes for use as **cartridge heaters,** and is wound around flat strips that are used as **band heaters.** Cartridge and band heaters, like those shown in figure 2.1, are used with temperature sensors and temperature controllers to control the temperature in automated moulding equipment and in soldering and heat-staking equipment.

A display provides information. While this may not be considered by a mechanically-oriented person to be "doing work," the display device is an actuator because it converts electrical energy to light energy.

Some display devices, such as a **Light Emitting Diode (LED),** are very simple devices. An LED is a diode that, when conducting DC current, gives off light. It needs very little current to work. LED **displays,** such as that shown in figure 2.2, consist of several LEDs, each of which is switched on or off by a separate output of a computer's parallel I/O port. At one time, most calculators came with LED displays, but low power consuming LCDs (discussed later) are preferred now. LED displays are still used if a bright display is needed so that it will be visible despite bright external light, such as in gasoline pump displays.

LEDs are used as the light sources in some light beam sensors (see chapter 4), in optical isolators (see chapter 7), and in fiber optic communication systems. In some such applications, it is necessary only to vary the *amount* of output light, and this can be accomplished by varying the voltage at the LED. It may also be necessary to vary the light signal's *frequency* in either of two ways: pulsing frequency or light frequency (color).

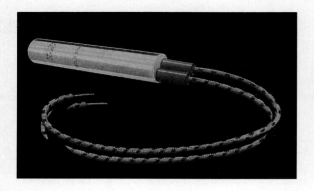

Fig. 2.1 Three types of heaters. (a) A cartridge heater; (b) a large diameter band heater; and (c) small diameter band heaters. (Photographs by permission, Watlow Electric Mfg. Co., St. Louis, Missouri.)

An LED light emitter can be turned on and off at a specially selected *pulsing frequency*. A matching light sensor, designed to respond only to light pulsing at the selected frequency, can be paired with the light emitter. The resultant light beam pair will not be sensitive to stray light from unintended sources.

Light frequency (color) variation can be accomplished only by using different types of LEDs. Each LED has a characteristic output color. Red, green, yellow, and the recently developed blue LEDs are most common. Fiber optic cables can now transmit many simultaneous communications if each signal is transmitted in a different color of light.

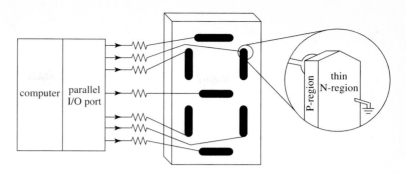

Fig. 2.2 An LED array for displaying numbers

Liquid Crystal Displays (LCD) have replaced many LED displays, because LCDs require much less current: only enough to create and maintain a charge between two small electrodes. LCDs do not generate light; they control the transmission of light. Figure 2.3 shows that many separate **liquid crystal cells** are used in a single liquid crystal display. Each liquid crystal cell has crystals that naturally tend to align parallel to each other horizontally. Each of the liquid crystal cells has a slightly different natural crystal orientation. The stack of liquid crystal cells is sandwiched between two **electrodes.** When the electrodes are charged, all of the liquid crystals move to a vertical orientation. Two **polarizing filters,** with polarization orientations 90 degrees to each other, are added top and bottom. Behind the layered LCD, there may be a **light source or a mirror.** The following explanation will assume a backlit display.

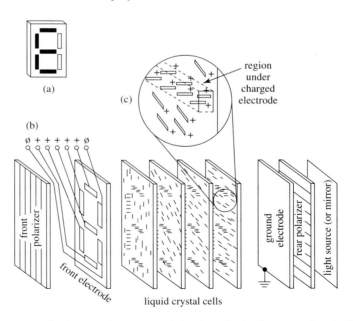

Fig. 2.3 An LCD display: (a) front view of display; (b) the layers in the display; (c) a liquid crystal cell

2.1 Light and Heat

When the LCD *electrodes are off:* Light from the light source enters at the back of the display. The rear polarizing layer allows light waves with only one orientation to enter. The electrodes are thin enough that light passes through them. Each layer of liquid crystal cell rotates the light waves slightly so that the light waves eventually rotate a full 90 degrees and can pass through the front polarizing layer.

When the *electrodes are charged,* the crystals between them stand upright. They cannot rotate the polarized light waves that enter through the back polarizer. The unrotated light waves cannot pass through the front polarizer, so they appear black on a light background.

Small LCD displays, such as in watches, have only a few electrodes. Large LCD displays (e.g., in laptop computers) require that many rows and columns of electrodes be controlled, and that they *allow* light to pass when charged. The polarizing layers must not be rotated relative to each other.

For color displays, three colors of backlight must be built in. Many LCD displays have a mirror behind the display instead of a light source; room light passes through the display and reflects back out if the electrodes are not charged.

Some early laptop computers were offered with **gas plasma displays** (typically red). This type of display is illustrated in figure 2.4. In these displays, charged electrodes cause trapped pockets of gas (often neon) to glow. Faster to turn on and off than LED or LCD elements, flatter than CRT displays, and easily digitally controlled, these displays were considered superior for portable computers. The drawback is that they are more expensive than LCDs and they draw more current, which means a portable battery will not last as long.

Where space and power consumption are not problems, the **Cathode Ray Tube (CRT)** display still predominates. CRTs are relatively cheap, and can be built sufficiently large and with adequate resolution for even Computer Aided Design applications. As shown in figure 2.5, a CRT tube has an electron gun that accelerates electrons in a narrow beam past deflection coils toward the center of the display surface. The amount of charge at each deflection coil controls the beam deflection, and therefore affects where it will hit the display surface. A fluorescent coating on the display surface glows as the electron beam passes through it to the thin transparent electrode that absorbs the electron beam.

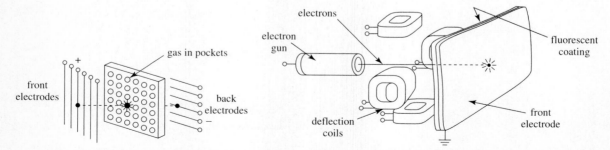

Fig. 2.4 Gas plasma display   **Fig. 2.5** The CRT display tube

In a standard television or computer monitor, the deflection coils cause the electron beam to sweep the display surface at a constant speed and in a repeated pattern called a **rastor scan.** The electron gun varies the beam's intensity during the rastor scan so that some portions of the display screen glow brighter than others. (For more detail see Vidicon Cameras in chapter 5.) For oscilloscope displays and CAD **graphics displays,** the electron beam does not follow a rastor scan pattern. The beam is deflected only to those portions of the display surface where it is required, and the beam's intensity may not be varied.

2.2 PIEZOELECTRIC FORCE GENERATORS

Piezoelectric force actuators don't generate much force. They are usually used only to move small amounts of air or fluids. They can, however, recycle quickly (to move a second bit of air) if supplied with a rapidly changing input signal.

Piezoelectric actuators are frequently used as **tone generators** and as **small speakers.** Piezoelectric materials have a crystalline structure that distorts when a voltage is applied across it. (It also responds to being distorted by developing a voltage, so it can be used as a sensor.) Very little power is needed to drive a piezoelectric speaker, because current does not move through it.

Another interesting application for piezoelectric materials is in high speed **ink jet printers.** The piezoelectric material in the orifice of the ink supply, as shown in figure 2.6, constricts when charge is applied. Electrically charged ink is expelled from the front of the orifice in a tiny droplet. While in the air, the droplet is deflected and accelerated by deflection coils so that it strikes the paper at a selected location.

2.3 SOLENOIDS AND TORQUE MOTORS

Where stronger forces must be exerted, an **electromagnet** can be used directly on a ferrous load, or can be used to push or pull a ferrous plunger against a nonferrous load.

A **solenoid** consists of an electromagnet and sometimes a ferrous plunger. Some solenoids and applications are shown in figure 2.7. Used as **relays,** small solenoids allow a low power circuit to move a **switch** controlling the current in a higher power circuit.

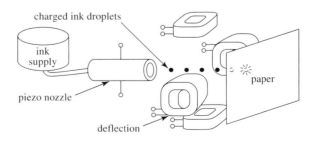

Fig. 2.6 Piezoelectrics in ink jet printers

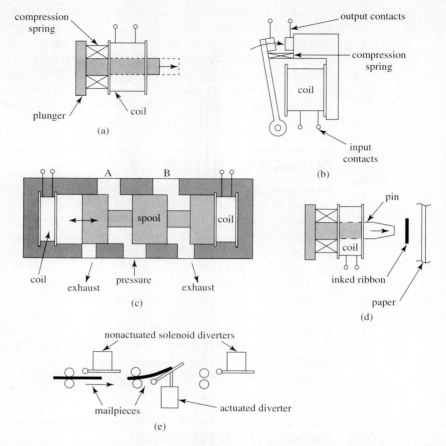

Fig. 2.7 Solenoids: (a) basic construction, (b) relay switch, (c) solenoid valve, (d) printer pin solenoid, (e) mail diverter

Some solenoids can move heavier loads, such as the spool in a solenoid-actuated pneumatic or hydraulic **valve.** The larger the coil, the longer it takes to actuate. Solenoids can be made to act more quickly to move light loads, such as pins in a **dot-matrix printer,** if larger power supplies are used to drive them. These solenoids must be built to withstand the high temperatures they generate. Even larger and slower solenoids are available to move heavier loads, such as a **diverter** in a mail sorting transport.

Solenoids, because they are electromagnets, do not exert the same force over their whole stroke. In figure 2.8 we see that some solenoids provide *more force at the end of their stroke,* while others provide more force at the *start* of their stroke. Solenoids can be purchased to *pull* or to *push* loads. Solenoids that have built-in *linear to rotary converters* are also available.

When selecting a solenoid, care must be taken with the **duty cycle** specification. "Duty cycle" describes how long the solenoid is held on during a machine cycle. To prevent them from burning out, solenoids for long duty cycle applications are built with heavier conductors. They are slower as a result. Where a long duty cycle solenoid is required, but the solenoid must be quick, control circuitry can be built to provide full power to move the load and plunger and then current can be reduced to hold the load.

Some **torque motors** contain solenoids. A torque motor is used to apply a rotational force but not a continuous rotation. The solenoids in a torque motor rotate an armature about its axis. If two coils are used as shown in figure 2.9, some of the non-linearity in the resultant torque can be eliminated. This type of torque motor has inherent limits on how far the rotor can be caused to rotate.

2.4 AIR-POWER ACTUATORS AND SOLENOID-ACTUATED VALVES

2.4.1 Actuators

The actuators described so far have all been electrically driven. There is also a wide range of pneumatically driven actuators, as shown by figure 2.10.

The actuator that is most frequently used in automation is pneumatically driven. This is the common **air cylinder.**

Air cylinders, shown in figure 2.11, are extended and retracted by compressed air. They are specified by stroke length and cylinder inner diameter. The diameter limits the force available. **Spring return** and **spring extend** cylinders are available, but **double-acting** cylinders are more frequently used, even though they require two air lines instead of one. The powered retraction feature means more dependable operation. Air cylinders

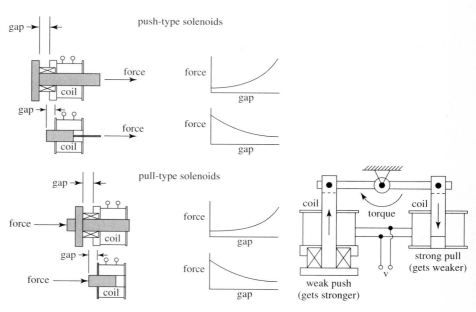

Fig. 2.8 Push and pull solenoids and force/displacement characteristics

Fig. 2.9 Electromagnetic torque motor

can be purchased with magnets in the piston heads, so that **reed switches** attached to the outside of the cylinder can detect piston extension or retraction.

The piston rod of an air cylinder is often **single-ended,** so the extension force is greater than the retracting force due to the difference in the air-bearing surfaces on the piston head. If a cylinder has a **double-ended** piston rod, extension and retraction forces are the same. Both ends of the piston rod may be used to move a load, or a sensor can be mounted on one end to detect piston rod position. **Non-rotating** piston rods are available, although delivery and seal life may not be as good.

Fig. 2.10 The Festo line of pneumatic actuators. (Photograph by permission, Festo Inc., Mississauga, Ontario, Canada.)

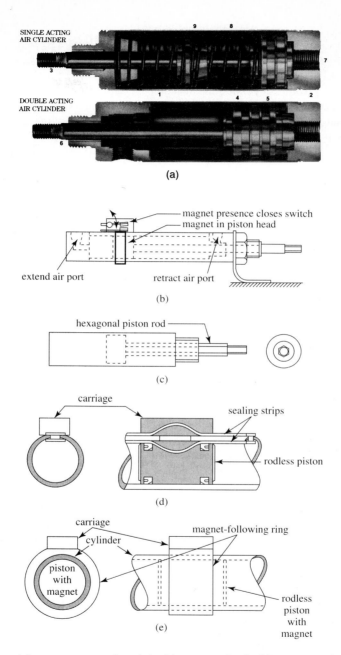

Fig. 2.11 Air cylinders: (a) spring retracted and double-acting; (b) double-acting with one reed switch; (c) double-ended, with non-rotating piston rod; (d) C-section rodless; (e) magnetic rodless. (Photographs by permission, Humphrey Products Company, Kalamazoo, Michigan.)

Cylinders can be purchased with an externally threaded cylinder head, or with a (usually rear) pivot mount, so that the cylinder can be rigidly fixed or allowed to pivot. Figure 2.12 demonstrates the range of cylinder-mounting options and the rod clevis attachments available.

Rodless cylinders are available for applications where space is limited. Some rodless cylinders (see figure 2.11 (d)) use a *C-shaped housing,* so that the piston head can be directly attached to a carriage riding along the outside of the cylinder. An ingenious multipart sealing strip prevents air from leaking out the open part of the C-section. These cylinders do not come in small bore sizes, due to the sealing strip geometry constraints. Another common rodless cylinder has a *piston head with a strong magnet* built in (see figure 2.11 and figure 2.13). A circular carriage surrounds the cylinder and follows the magnet. Coupling between the piston head and carriage is remarkably strong.

Rotary actuators are also available for use with compressed air. They typically consist of an air chamber containing a rotor that can be forced to stops in either direction. These compact air-powered rotary actuators can generate only low torque. If more torque is required, it can be attained inexpensively by using an air cylinder to push an arm about a pivot.

Air is compressible, which can lead to sluggishness and unpredictability problems. Pistons may stick when they should move, so air pressure builds up until they jump. **Lubrication** in a compressed air delivery system can help solve the problem, but lubricant

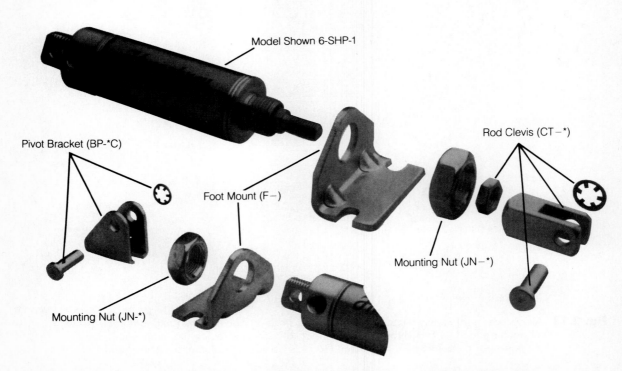

Fig. 2.12 Cylinder mounting and rod clevis options. (Photograph by permission, Humphrey Products Company, Kalamazoo, Michigan.)

Fig. 2.13 Rodless cylinder, magnetic-type. (Photograph by permission, Humphrey Products Company, Kalamazoo, Michigan.)

delivery is never perfect. The problem becomes worse with large seal-bearing surfaces, so a good compromise is to select cylinders with the minimum bore, and then drive them with full available air pressure. One easily-corrected reason why cylinders may jump is that their air supply lines may be too small or are constricted, so air pressure builds up too slowly in the cylinder. If jumping is a problem or a safety hazard, the best solution is to *use a flow control valve on the outlet* port of the cylinder to limit the speed of the cylinder and provide lots of air pressure at the inlet port.

Other air-driven actuators include several types of **vane and piston air motors,** used where sparks or electromagnetic interference would make electric motors undesirable.

Venturi vacuum generators are small and light enough that they can be mounted right on robotic vacuum-cup grippers. Positive air pressure causes air flow through the venturi, which draws air through the port attached to the vacuum cups. Many venturi vacuum pumps provide a puff of positive air pressure at the vacuum port when they turn off, which helps in releasing loads held by vacuum cups.

Dispensers for glue, solder paste, and similar fluids are commercially available from some glue or solder manufacturers. These mechanisms use controlled air pressure to force well-metered quantities of the fluids through needles every time a button is pushed, or when a computer controller demands the dispenser to work.

Air-powered tools of all sorts, from hand-held **screwdrivers,** to vehicle-mounted **rockdrills,** to bench mounted **presses,** are available, and all can be easily integrated into automated manufacturing systems.

2.4.2 Valves

When an air-powered actuator is used in an electronically controlled system, the controller's output signals (usually DC) have to control air flow. Conversion of DC signals to pneumatic pressure or flow requires the use of a **solenoid-controlled valve.**

There are two types of simple air valves. One type, shown in figure 2.14, p. 32, is the **poppet** valve, which is an inexpensive valve for controlling air flow in a single direction through a single line. Such a valve would be adequate for controlling a venturi vacuum generator.

The other type, the **spool** valve shown in figure 2.14 (continued), pp. 33 and 34, is usually purchased with three or with four ports for connecting air lines. One port is for pressurized air, another is for exhaust air, and the remainders are for connecting actuators. The spools may be spring-return or detent.

Most two- and three-way valves have **spring returns.** The spring returns the spool to its normal position when the solenoid is not powered. Two-way valves open and close the air path between two ports. Three-way valves connect one air port to either of two other ports. A three-way valve could switch an actuator between supply air and an exhaust port. The actuator could be a spring-return cylinder. Spring-return three-way spool valves can be used as pressure-release safety valves; the valve must be powered to connect the air supply to the system, and if power fails it exhausts system pressure.

Four-way valves can be either spring-return or detent. Four-way valves control two output ports. In one valve position, output port A is connected to the air supply and port B is connected to exhaust. In the other position, B receives supply air while A exhausts. This type of valve is required for double-acting cylinders, because the piston must be driven both ways. **Spring-return** valves are chosen where the air-powered actuator has a position it should return to if power fails, or if the actuator is driven only for short intervals. **Detent**

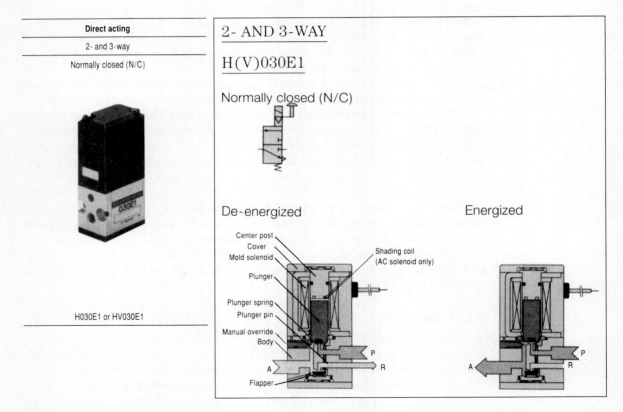

Fig. 2.14 Air control valves: direct-acting ("poppet"). (Photographs and diagrams by permission, Humphrey Products Company, Kalamazoo, Michigan.) Continued on next page.

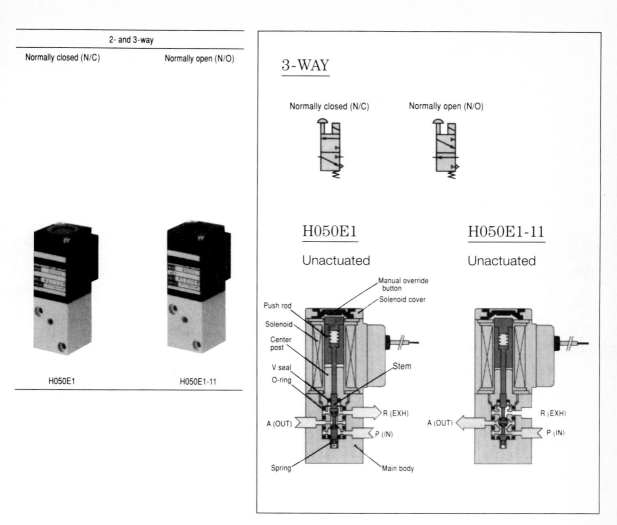

Fig. 2.14 Air control valves: spring-return two- and three-way. (Photographs and diagrams by permission, Humphrey Products Company, Kalamazoo, Michigan.) Continued on p. 34.

valves hold themselves in the position to which they were last solenoid-driven. They require two solenoids, one for each direction, but each solenoid needs to be only briefly actuated to move the spool. They stay cooler and use less power. Detent valves should be used if safety requires that the air-powered actuator not move if electrical power fails.

2.4.3 Air Supply

Air-powered actuators require a source of compressed air or vacuum. Many manufacturing plants have **compressors** and a ready supply of compressed air. Experienced users of compressed air are aware that the air must be *filtered* to remove dirt that can jam the actuator or valve, *regulated* to not exceed the design pressure, and *lubricated* with a fine oil

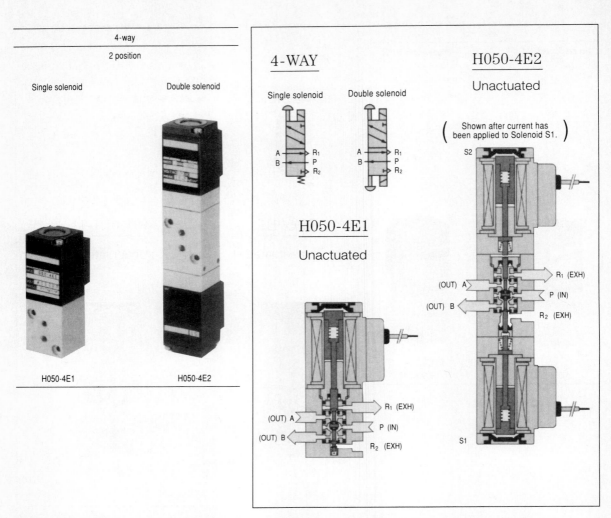

Fig. 2.14 Air control valves: spring-return and double solenoid ("detent") four-way. (Photographs and diagrams by permission, Humphrey Products Company, Kalamazoo, Michigan.)

mist to reduce wear in the actuators. Complete filter-regulator-lubricator (**FRL**) units (see figure 2.15) are available.

The presence of an FRL does not necessarily mean that lubrication is adequate. In an automated system, where air is always being turned on and off, oil may settle in the supply lines. Some pneumatic actuators may suddenly become hydraulic actuators when they fill with oil, while other actuators may never receive any lubrication. It is important that lubricators be located as near as possible to the actual inlets of actuators.

For a **vacuum supply,** the previously-discussed **venturi vacuum generator** can supply high vacuum pressure at low volumes. If high vacuum pressure and large air volume are both required, **vacuum pumps** are available. If only a low vacuum pressure and lots of air flow are required, then a **blower** or **regenerative blower** is a better choice.

2.5 HYDRAULIC ACTUATORS AND VALVES

Hydraulic actuator and valve choices are very similar to pneumatic choices, except that the actuators are more powerful, faster, and must be more robustly built. There is a much wider choice of hydraulic *rotary* actuators than is the case with pneumatically powered actuators, because the high pressures used in hydraulic systems allow them to develop reasonable torque.

Since hydraulic oil (and the water-based substitutes that are sometimes used) are relatively incompressible, position control is possible. A well-designed hydraulic system is not subject to the same jerky actuation that is sometimes seen in pneumatic systems.

For position control, special types of hydraulic valves are often used. The servovalve and proportional valve (see figure 2.16) are both spool valves that open proportionally with analog DC.

The several types of **true servovalves** all include internal servomechanisms to ensure precise proportionality. An electric input signal causes a torque motor to move a spool-control mechanism that provides pressure to move the spool. The feedback link readjusts the spool-control mechanism as the spool reaches its desired position.

Feedback links do not have to be mechanical as in this simple diagram; many are hydraulic.

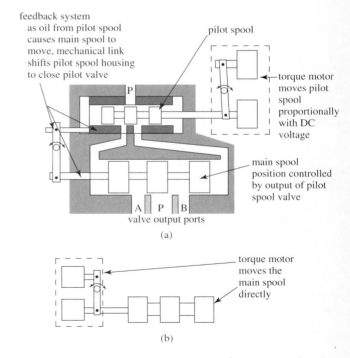

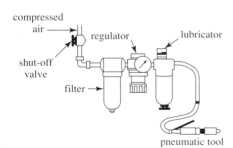

Fig. 2.15 Filter-regulator-lubricators in a compressed air supply system

Fig. 2.16 Servovalves: (a) one type of true servovalve; (b) proportional valve

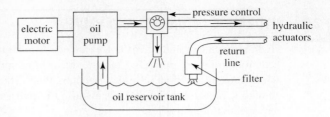

Fig. 2.17 Hydraulic power supply

Proportional servovalve spools are moved directly by torque motors, roughly proportionally with the input signal. Proportional servovalves have improved over recent years so that many now come with spool position sensors and external controllers that ensure their output is as proportional to input as a true servovalve's would be.

Hydraulic **power supply units,** such as that shown in figure 2.17, must include a *pump,* a *pressure control valve,* an *oil reservoir tank,* and a *filter* to clean the hydraulic fluid as it returns. If precision components such as servovalves are used, the fluid in the tank should be agitated and cleaned continuously.

2.6 CONVEYORS, FEEDERS, AND INDEXING TABLES

Several devices used in automated systems to transmit forces or to move workpieces and products are described here. The reader should recognize that this section is only a brief survey of such devices, and that many common options are not mentioned.

Non-stretch conveyor belts are commonly used to transport workpieces. **Stretchable belts** can be used to sandwich parts being moved, to protect them from damage. All belts stretch, but "non-stretch" belts do not stretch as much as "stretch" belts when put under tension. Both require crowned or flanged pulleys if they are not to ride off the pulleys. Conveyor belts are designed to have either a high ("no-slip") or a low coefficient of friction.

Modern flexible manufacturing systems often use parallel-belt conveyors to transport pallets from one workcell to the next. Each workcell performs work on the part(s) that are carried on the pallets. The pallets are lifted from the belt conveyor at each workcell and held in precise locations while work is being performed on the pallet's contents. Figure 2.18 shows a pallet and the pallet-locating mechanism at a workcell.

Antistatic synthetic conveyor materials are available and are used where static discharges can damage the product being conveyed. Metal conveyer belts do the same and are more durable, but more expensive and awkward to work with.

Chain conveyors can be constructed using standard roller chain or one of the special chain products shown in figure 2.19.

Plastic chains are often used to convey products or pallets containing products. These chains have the advantages that they are sterilizable for food handling operations, do not corrode, are light and cheap, and have a lower co-efficient of friction than many belts or metal chains. Some frequently-used plastic chains can **multi-flex,** which means they can carry products around bends as well as wrap around vertical pulleys. Figure 2.20 shows a multi-flexing, low friction plastic chain that can be used to move pallets containing a product from one workcell to the next in an automated manufacturing system. The pallet must be lifted from the chain at each workcell by a pallet lift and locate mechanism, because the chain does not stop running.

Plastic chains with free-running rollers at each link are available so that the chain can continue to run under stationary products that are not lifted from the belt. The rollers prevent tension from building up in the chain, and prevent products accumulated at a stop from exerting too much pressure on products ahead at the stop. These chains are known as **low back-pressure** or **no-back-pressure chains.**

Often, loose parts must be oriented as they are delivered to automated equipment. **Vibratory feeders** can be used. The operation of a **linear vibratory feeder** is shown in figure 2.21. A rail, supporting the product, is attached to angled metal leaf springs. When the electromagnets go on, the springs pull the rail backward and down so fast that the product is left floating in air. As the product begins to fall, the electromagnets release the springs, and the rail springs up and forward. The product on the rail is thus tapped forward. The amplitude, and sometimes the frequency, of the AC that drives the electromagnets can be adjusted to optimize the rate at which the product vibrates forward. Vibratory feeders can even cause products to ride uphill along the rail.

(a)

(b)

Fig. 2.18 Pallet entering a workcell (a), and then lifting and locating (b). (Photograph of pallet lock by permission, Bosch Automation Products, Buchanan, Michigan. Photographed in automated system by permission, Automation Tooling Systems, Inc., Kitchener, Ontario, Canada.)

1700
width:
2-11/64"

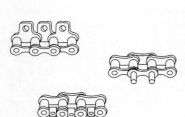

Fig. 2.19 Special roller chain options

Fig. 2.20 Multi-flexing chain, pallet, and precision pallet holder. (Photograph of chain by permission, Rexnord Corporation, TableTop Operations, Milwaukee, Wisconsin.)

Rotary vibratory feeders, usually called **bowl feeders,** support a metal bowl that rotates as it rises and lowers. The workpieces vibrate up ramps around the perimeter of the bowl. Bowl feeders are custom-built for the product they are to feed, and diverters are built into the ramps around the perimeter so that workpieces that are not correctly aligned or are improperly separated tip off the ramp back into the bowl.

Bowl feeders are noisy. They can be coated with a urethane to dampen some of the noise, or may be totally enclosed in noise absorbing housings. Some products may be damaged by falling and vibrating against their neighbors. Bowl feeders work best when only lightly filled, so another device is often required to replace the bowl's contents it empties. Despite these drawbacks, bowl feeders are frequently used in automated systems.

Indexers are used to move a conveyor or work table such that, as a motor continuously rotates the indexer's input shaft, the indexer outputs intermittent motion, with precisely controlled stopped position and stopped time (called "dwell" time). **Rotary indexers** can be used to rotate circular indexing tables around a ring of workstations, or to move an indexing chain linearly along a sequence of stations. A rotary indexing table is shown in figure 2.22. Indexer motors can be allowed to run continuously, so that each dwell always takes the same time, or can be controlled by a digital controller that makes the indexer move only after all workstations have signaled that they have completed their work. Indexers are ordered by specifying the number of stops per revolution, the rotation speed, and the dwell time in continuous running mode. Indexers are then custom-assembled to specification, so delivery time is usually long and the price is high, but the precision and dependability make the wait and cost worthwhile.

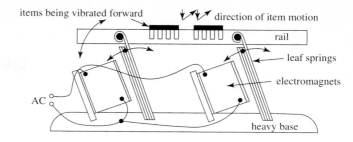

Fig. 2.21 Vibratory feeders

Small **single-turn indexers** are cheaper and can be used for light loads where precision is not as critical. They have a small electric motor that rotates the output shaft when a clutch is engaged, and a solenoid-actuated ratchet mechanism that releases the clutch to engage the output shaft for exactly 360 degrees. Variations are available for rotations of less than 360 degrees.

2.7 POWER TRANSMISSION COMPONENTS

Power transmission components deliver power from an actuator to where it is needed. They may also convert between rotational power and linear power in the process, and may exchange torque for force or rotational or linear speed.

Since

$$\text{Power} = \text{Torque} * \text{Rotational Speed} = \text{Force} * \text{Linear Speed}$$

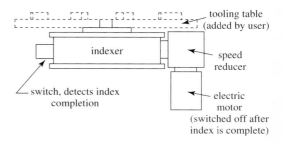

Fig. 2.22 Rotary indexing table. (Photograph by permission, Fusion Inc., Willoughby, Ohio.)

 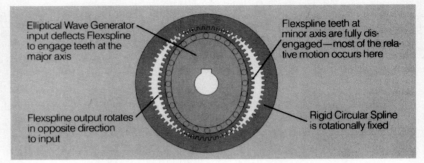

Fig. 2.23 Harmonic drive gear reducer. (Photograph and diagram by permission, Harmonic Drive Technologies, Teijin Seiki Boston Inc., Peabody, Massachusetts.)

then, if a chain is used as a power transmission device to carry power from an electric motor, the chain converts the motor's torque to linear chain speed. If the motor's chain-drive sprocket is increased in diameter, the chain would be driven faster, but with less force.

Gears are sometimes used to trade off rotational speed for output torque. Fully assembled gear reducers are available, and are common enough that they will not be discussed here, with the special exception of harmonic drives.

The **harmonic drive,** shown in figure 2.23, is a gear unit used extensively in devices such as robots, where large gear reductions are necessary. This device has unusually low power losses, and weighs much less than conventional gear reducers. On the other hand, the harmonic drive costs more and requires more careful design of mating parts.

The harmonic drive's *eccentric-shaped "wave generator"* can rotate inside the *inner gear ring (called the "flex spline")*. Roller bearings separate these components so they do not wear on each other. The flex spline is distorted by the wave generator so that only two points on its ring of external teeth engage the inner teeth of the *non-rotating outer gear ring (called the "circular spline")*. As the power source rotates the wave generator one full revolution inside the flex spline, it causes the locations of the contact points to rotate one full revolution. Since the flex spline has fewer teeth than the circular spline (e.g., 118 inner:120 outer), then the *flex spline will be caused to rotate* only a small amount (e.g., 2 teeth out of 120), and gear reductions of the order of 60:1 are typical.

Timing belts have teeth molded into (usually) one side. These teeth engage with teeth in drive and driven pulleys, so slip does not occur between the belt and pulleys. If the inertia of the load is small enough and if acceleration is not too great, timing belts can even be used in such precision applications as the control of the position of print heads in printers.

To move a carriage to precision locations within a restricted linear range, stepper motors or servomotors (see chapter 3) are sometimes used to drive the carriage via a **ballscrew and ballnut.** The ballscrew and nut are similar to a standard threaded rod and nut in that, if the screw turns and the nut is not allowed to, then the nut must move along the screw. As shown in figure 2.24, ball bearings between the screw and the nut allow precision fits, with very little backlash and much less friction.

2.8 SPECIAL-PURPOSE ACTUATOR SYSTEMS

Some devices used as actuators are complete automated systems in themselves. Robots are a prime example. We will leave robots to another chapter, where we will examine them as examples of automated systems, but we will look at two of the modern robot's ancestors here.

Off-the-shelf **X-Y tables** are used as components in many systems where reprogrammable position control is desired. The X-Y table, as shown in figure 2.25, is just two servocontrolled linear actuators (often servomotors and ballscrews) with their linear axes perpendicular to each other. The X-Y table may move the table under a stationary tool (such as a glue dispenser), or may move a tool over a stationary worktable. Programmable X-Y tables are used to control **water-jet cutters** that are used to cut shoe leather, plastics, and floor mats for cars. Clothing manufacturers are improving efficiency by using powerful X-Y tables to move knives or lasers used to cut materials. The material cutter in figure 2.26 can simultaneously cut many layers of material, and a computer optimizes the nesting of components to reduce waste. Even the woodworking industry is getting into the act. X-Y tables are used in automatic **panel cutters** to saw computer-optimized patterns from four by eight feet sheets of plywood and veneer.

Numerical control (NC) equipment can cut metal under the control of a built-in digital controller instead of being operated by a human operator.

A piece of NC equipment combines X-Y motion of a workpiece or cutting tool with a third, the Z axis. In addition to linear position control, control of rotation about each of the three axes is available with some NC equipment. Additional controllable features include the rotational speed of the tool or workpiece and the flow of coolant.

Early NC lathes and mills required a program of instructions to be prepared as patterns of holes punched into paper or plastic tape. The tape was read one instruction at a

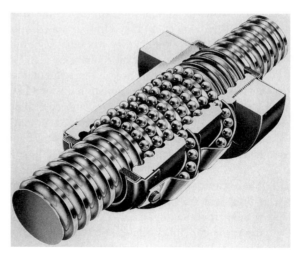

Fig. 2.24 Ballscrew and ballnut. (Photograph by permission, Warner Electric/Dana Corp., South Beliot, Illinois.)

Fig. 2.25 An X-Y table. (Photograph by permission, Panasonic Factory Automation Company, Franklin Park, Illinois.)

time while the workpiece was machined. The program tapes were typed using equipment specifically designed to punch tape.

Later, read/write memory was added in NC controllers, so the tape had to be read only once into memory. The NC equipment became known as **Computer Numerical Control (CNC)** equipment. Meanwhile, advances in tape punchers allowed programmers to write whole NC programs at a standard computer terminal, then send the program to a tape puncher.

Logically, the tape was eliminated in the next evolution by having the central computer write programs directly into CNC memory. The CNC equipment became known as **Direct Numerical Control (DNC)** equipment.

Now the human programmer is no longer always required. **Computer Aided Manufacturing (CAM)** systems are now available that receive part descriptions from a CAD database, automatically write the NC program, and can even provide instructions to the NC equipment as it cuts the metal.

There are **systems houses** that will build special-purpose automated equipment units of any kind. Each of these custom-designed systems can be used as a component in a larger system, with the larger system's controller controlling the sub-systems' controllers.

This chapter on actuators will never end if we include custom equipment. There are, however, some types of special-purpose equipment that are built in sufficient quantity and with sufficient standard features that they are worth mentioning.

Some special-purpose systems assemble electrical components onto circuit boards. These **component insertion machines** can be purchased to handle any of the electronic component shapes. A typical component insertion machine might remove components from their packaging in a roll, cut and bend the leads to the proper length and shape, then insert the component into the correct location on the board. Programmable component insertion machines can select components and positions.

Some systems convey assembled circuit boards through **component soldering machines,** where solder is applied or where pre-applied solder on the boards is heated

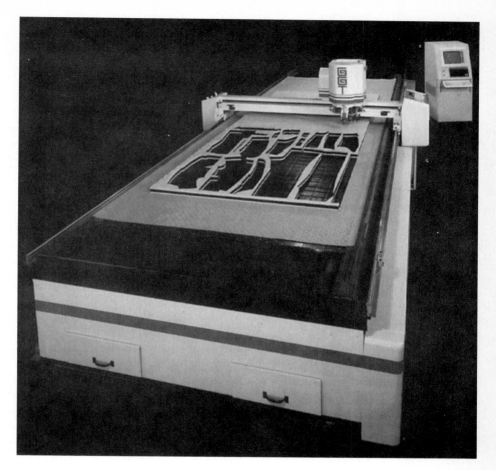

Fig. 2.26 An X-Y table for cutting material for clothing. (Photograph by permission, Gerber Garment Technologies, Tolland, Connecticut.)

until it flows to attach the components and then transports the circuit boards through **board cleaning machines** where solder flux is washed off.

Completed circuit boards are often tested by **automated circuit and component testers,** in which the boards are drawn down onto an array of probes, and the tester automatically runs through a series of tests, notifying the human or computer controller if a fault is detected. Chapter 9 describes an automation system that includes a circuit board tester.

Automated Storage and Retrieval Systems (AS/RS) can be used for computerized control of inventory. A typical AS/RS includes an automated method of moving goods from a central shipping/receiving location into and out of storage. The AS/RS will keep track of exactly what is in stock and where it is. This means that stock items do not need to be stored in the same location every time, as long as the computer remembers where they are. More efficient use can be made of storage space. The computer also can keep track of stock reserved for outstanding customer orders, without actually having to pull it from storage, until the whole order can be filled. The computer can ensure that the reserved items are not given to anybody else.

The AS/RS is a valuable component in a fully automated plant. It can be used in Computer Integrated Manufacturing (CIM) systems that accept customer orders, schedule production, produce and assemble products, assemble and ship customer shipments, and even invoice customers. Control of partial orders as goods move into and out of inventory is essential for such a CIM system to work.

We have already moved somewhat beyond what most people would consider as simple actuators. Before we move on to discussions of systems that use actuators, the following chapters will discuss motors and motor control, sensor technology, and the controllers required to integrate actuators and sensors into automated manufacturing systems.

2.9 SUMMARY

Actuators perform work. When provided with power of one sort (e.g., electrical), they output power of another sort (e.g., force).

Heaters and displays are actuators. Resistance heaters generate heat. Light emitting diodes (LEDs), gas plasma displays, and cathode ray tubes (CRTs) generate light. Additional control circuitry is often required to select where the display elements should glow. The liquid crystal display (LCD) uses little power because it doesn't have to generate light, just control the display elements so that they control where light can pass.

Electric energy can be converted to small forces by **piezoelectric** materials or to larger forces by **solenoids.** Solenoids are often used directly as the actuators, where the electromagnetic force moves a machine component. Solenoids are also extensively used as the force sources inside hydraulic and pneumatic solenoid valves. The solenoids open or close the hydraulic fluid or air ports and the valves deliver fluid to move the final actuator. **Torque motors** are solenoids used such that they act on levers to generate a torque proportional to the DC power they receive.

The most common **fluidic actuators** are **cylinders,** which are available in a wide variety of bore sizes and lengths. Options include special rod shapes, double-ended rods, assorted mounting styles, reed switches, spring-return or double acting, and even a few types of rodless models. **Rotary actuators** are also available. Some rotate less than 360 degrees, but others, often called fluid motors, can rotate without limits. Fluidic actuators need valves. A **valve** may be three-way, if it has only three ports, or four-way if it has separate ports for each of the supply and return fluid and for the two ports of the actuator. The valve may be a poppet or a spool valve. Spools can be spring-return or detent. **Servovalves** open proportional to the DC they receive. Pneumatic actuators require filtered, pressure-regulated, and lubricated air. Flow control valves can improve pneumatic actuator operation. Hydraulic actuators require a hydraulic pump, fluid reservoir, filtering, pressure control, and sometimes oil cooling.

Products can be moved if the actuator moves **belts, chains** (including plastic chains), **ball screws, vibratory feeders,** or **indexers.** A power transmission medium may be required to change speed, torque, or force. Gear reducer units are also available for this purpose. **Harmonic drives** are gear reduction units for large speed reduction with little power loss and where light weight is desirable.

X-Y tables and **NC/CNC/DNC** equipment are automated systems that can be treated as actuators.

Systems houses will assemble automated systems or sub-systems. Demand is sufficient that some automated systems, particularly for the assembly of electronic circuit boards, are available as standard equipment. **Automated Storage and Retrieval Systems (AS/RS)** are computer-controlled actuators that make control of inventory possible in a Computer Integrated Manufacturing (CIM) system.

REVIEW QUESTIONS

1. Describe an application in which:
   a) an LED display would be preferable to an LCD display.
   b) an LCD display would be preferable to an LED display.
2. How does a rastor scan CRT display operate?
3. List four different applications for solenoids.
4. Why does a solenoid not output the same force over its full displacement?
5. If you where using a double acting air cylinder, and if the cylinder was to be extended for about half the time it was in use, which type of spool valve would you select and why?
6. If the air cylinder in question 5 was required to extend slowly at a very constant speed, where would you install the flow control valve?
7. What device should you install in the air-supply line of any pneumatic actuator? Why?
8. How does a proportional (hydraulic) servovalve differ from a true servovalve?
9. What is meant by the term *low back-pressure* as it is used to describe a conveyor chain?
10. How does a vibratory feeder cause material to move?
11. What are rotary indexers used for?
12. In what ways is a harmonic (gear) drive better than conventional gearing?
13. How does DNC differ from CNC? What does a three axis CNC lathe need to control that an X-Y table doesn't need to control?
14. What is an AS/RS and what does it do?

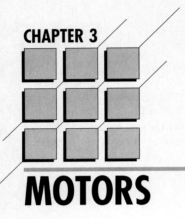

CHAPTER 3

MOTORS

CHAPTER OBJECTIVES

After completing this chapter, the reader will be able to:

▬▬▬ Discuss in general the construction and principles of operation of electric motors.

▬▬▬ Recognize the differences in construction and operation of the following:

DC motors:
- permanent magnet motors
- brushless DC motors
- series, shunt, and compound wound field motors

AC motors:
- induction motors
- synchronous motors
- universal motors
- single phase motors of various types

Electronically commutated motors:
- stepper motors
- timing motors
- SCR motors
- brushless DC servomotors

▬▬▬ Specify and use the following techniques for controlling:

DC motors:
- armature voltage control and field voltage control
- dynamic braking and plugging
- DC pulse width modulation

AC motors:
- AC pulse width modulation
- variable voltage inversion
- cycloconverter

Electronically commutated motors:
- SCR control
- stepper motor control

3.1 THE HISTORY OF ELECTRIC MOTORS

The development of motor types and the development of techniques to control and transmit electricity have gone hand-in-hand. The earliest electrical power companies, like the Edison Electric Company, provided direct current (DC) power from many small generating plants. DC power can't be transmitted very far, due to resistive losses in the wires, so these plants could supply only about one square mile each. DC motors became popular at this time. They are generally controlled by changing the DC voltage or current supply.

In the early 1900s, two developments promoted the use of alternating current (AC). First was the building of large generating plants, such as the Niagara Falls hydroelectric plant, which made it necessary to deliver power further. AC power was used because it doesn't suffer from line losses as badly as does DC.

The second factor was the development of motors that would run on AC. These early AC motors ran at one speed only, defined by the frequency of the AC. Controlled

Fig. 3.1 Motors and motor controllers. (Photograph by permission, Parker Hannifin Corporation, Compumotor Division, Rohnert Park, California.)

speed or torque applications were left to DC motors. Since the world had now adopted AC power, the most practical way of acquiring DC power for many years was by a roundabout method: The supplied AC power would be used to drive an AC motor turning a DC generator, and the generator's output could then drive the DC motor!

More recently, the electronics revolution has given us the ability to alter standard 60 Hertz AC to different frequencies, to different wave forms, and even to convert it into DC. These developments have spawned new breeds of motor: electronically commutated motors, stepper motors, and Silicon Controlled Rectifier (SCR) motors. These motors are often classed as DC motors, but some of them are more similar in construction and operation to AC motors. We will refer to these motor types as "electronically commutated" in this chapter.

3.2 CONSTRUCTION OF ELECTRIC MOTORS

Electric motors, being devices to convert electricity to rotary motion, must all be constructed alike to a certain extent. Figure 3.2 shows the construction and some of the terminology used in describing the construction of various types of electric motors.

All electric motors have a central rotating section called a **rotor** or an **armature.** A rotor usually does not require connection to the power supply. It may contain conductors, permanent magnets, or alloy metals selected for their magnetic properties. Some rotors have copper windings that are connected to the current supply through **slip rings.** An armature is similar to a rotor except that it includes windings of copper wire to which electric current is supplied via a **commutator.**

Slip rings are simple electric connections. The inner rings of the slip ring pair can rotate inside the stationary outer rings without loss of the electrical connection. The slip rings do not affect the direction of current flow through the rotor. A commutator, on the other hand, is intended to control the direction that current flows through the armature. A motor with an armature also includes carbon brushes, mounted in the housing, that conduct electricity to the portions of the rotating commutator that they are bearing on at that time. We will see why control of current direction in the armature is important later, in the section on theory of operation of electric motors. Some motors have neither slip rings nor commutators.

The rotor or armature is supported on **bearings** in the housing of the motor. Radial roller element bearings are common, and these may require periodic lubrication on larger motors. Many very small motors now come with oil-filled bronze bushings instead of roller bearings. Some heavy duty motors come with thrust bearings as well as radial bearings.

The stationary housing of the motor (hence *stator*) provides the magnetic **field** essential to the operation of the motor. The field may be provided by permanent magnets or by electromagnets. If the field is electromagnetic, the field strength can be varied. In some motors, the orientation of the magnetic field can be controlled. Later sections explain the effect of the field.

All electric motors require electrical power to operate. Some motors need only to be connected to 120 Volt, 60 Hz AC. Others require only simple DC power supplies. More sophisticated power supplies, often called **motor controllers,** are needed if:

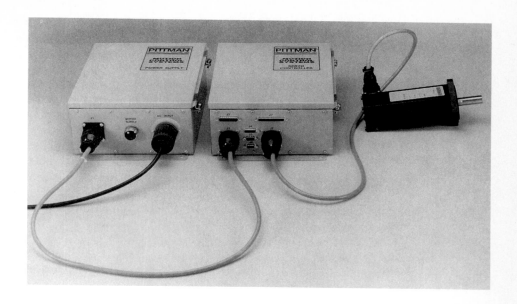

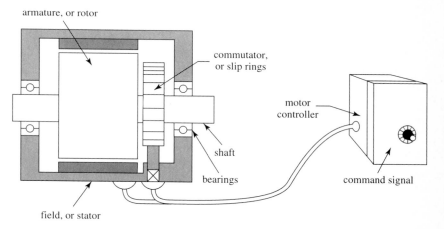

Fig. 3.2 Construction of an electric motor. (Photograph by permission, Pittman, A Division of Penn Engineering & Manufacturing Corp., Harleysville, Pennsylvania.)

- The motor is large, and it requires gradual starting or stopping to avoid damaging surges of current.
- The motor torque, speed, or shaft position must be controlled.

Motor controllers, usually purchased with the motor, may convert AC to DC. They may allow control of DC power, or of AC frequency, or may simply consist of switches that open and close in a timed sequence.

In an automated system, a **command signal** to the motor controller can be used to specify the motor speed or shaft position. The command signal may be an analog DC signal from a PLC (programmable controller) or robot controller in the same workcell as the motor, or from a cell controller in the larger Flexible Manufacturing System. In a Computer Integrated Manufacturing (CIM) system, the command signal might originate with the plant host computer and arrive as a digital value via a Local Area Network (LAN). The motor controller would require a communications interface to receive LAN messages.

Figure 3.2 does not show some common motor components. Some motors, specifically those used as timing motors, may have mechanical *speed governors* built in. These motors are gradually being replaced by timing motors that have their speed controlled electronically. Electronically-timed motors may have special purpose circuit boards in the rotor or the housing, and may include **extra magnets** or **wiping contacts** for control purposes.

Some motors, specifically single phase AC motors, require **capacitors** or **inductors**, and sometimes have extra electrical connections to the rotor that disconnect these components once the motor has started. We will examine motors of this type in a later section of this chapter.

Motors for use in speed or position control applications can be purchased with built-in **speed or position sensors.** These motors, called servomotors, are often supplied with especially **light armatures** so that they can be quickly accelerated on demand. Brushless DC servomotors (discussed later) are sometimes built with **stationary central armatures,** and sometimes built *inside out,* with a rotating armature surrounding a stationary field.

3.3 THEORY OF OPERATION OF ELECTRIC MOTORS

All electric motors work through the interaction of electric current, magnetic fields, and motion or torque. Three principles of operation can be isolated to define the operation of motors in simple terms:

1. Opposite magnetic poles attract each other.

Most motors do not operate using this principle! Several motors do, however, the most important being **AC synchronous motors** and **stepper motors.** These motors have rotors with permanent magnets (or materials that when magnetized have a "reluctance" to lose magnetic orientation) so that the rotors can be caused to exactly follow magnetic fields as the fields rotate. We will examine how fields are caused to rotate in later sections of this chapter.

2. A current moving through a wire across a magnetic field will cause a force to be exerted on the wire.

It can be shown that this principle is only a result of the first principle, since current moving through a wire causes a magnetic field to be set up around that wire, and the interaction of that magnetic field with the larger magnetic field causes a force. This principle, called the "Left Hand Rule," is demonstrated in figure 3.3. The first finger points in the direction of the magnetic field from North to South. The second finger points in the direction that (conventional) current is flowing through a wire. The thumb points in the direction that the wire is being pushed by the resultant force.

Most **DC motors** operate on this principle. Their speed can be controlled by altering either the amount of current flowing through the wiring in the armature, or by adjusting the

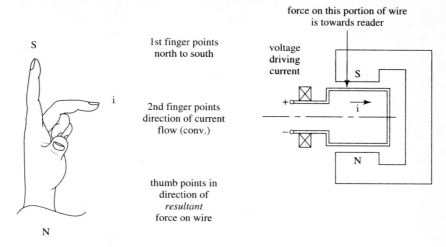

Fig. 3.3 Left hand rule

surrounding electromagnetic field strength. The effect of the third principle cannot be ignored in applying this second principle.

3. When a conductor moves across a magnetic field, a voltage is produced in it that will cause a current to flow (if the conductor is part of a closed circuit).

This principle is the one employed to generate electricity from sources such as falling water or steam from the burning of oil or radioactive decay. These energy sources turn the armature of a generator in a magnetic field, causing current to flow in the windings in the armature, and from there, through the electric power distribution system.

The same principle tends to cause current to flow in the windings of motors as they rotate inside their surrounding magnetic fields.

Let's look at the effect of this principle on the DC motors we examined in discussing principle 2. We saw that current moving in a magnetic field was the source of torque in DC motors. Principle 3, described using the Right Hand Rule in figure 3.4, causes most DC motors to settle at a speed roughly proportional to the DC voltage supplied to the armature windings. The first finger points along the magnetic field, from North to South, as in the Left Hand Rule. The thumb now indicates the direction that the conductor is being moved relative to the field. The second finger now points in the direction that the resultant "electromagnetic" force (voltage) is trying to cause current to flow in the conductor.

Note in figure 3.4 that the induced voltage, often called the **Counter-Electro-Motive Force,** or CEMF, is acting in the opposite direction to the voltage that caused the current that led to the DC motor turning in the first place! So the faster the motor rotates, the greater the CEMF. The motor speed will settle at where the balance between applied voltage and CEMF allows just enough current flow to overcome the load that the motor is turning. It should be apparent that adjusting either the magnetic field strength or the applied voltage that causes current in the armature will result in changes in the motor torque and therefore in motor speed. We will discuss these motor control techniques later.

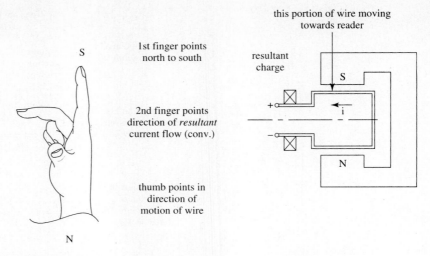

Fig. 3.4 Right hand rule

AC induction motors also make use of these second and third principles. This motor is called "induction" because, as an AC-generated field rotates around a stationary rotor, the *relative* movement of stationary conductors to the moving field causes current to be induced in these conductors (right hand rule). The presence of current in the field causes the rotor to rotate (left hand rule). When the rotor speed is slightly less than the speed of rotation of the field, the induced current and torque is just sufficient to move the load attached to the rotor. AC induction motor speed can be controlled by controlling the speed of rotation of the field. Since no brushes or slip rings are needed to cause current to flow in the rotor, this type of motor is sometimes called "brushless."

TYPES OF ELECTRIC MOTORS

The three major types of electric motors include DC motors, AC motors, and the motors we are calling electronically commutated motors (usually classified as DC motors).

We have already mentioned some specific sub-classifications of these three types of motors, and indeed, there is a bewildering array of motor types offered by manufacturers, who may or may not give a sufficient description of their product to allow the buyer to determine which type of motor he or she is being offered. Figure 3.5 is an attempt to classify the more popular motor types.

Some of the motors in figure 3.5 have names describing where they are typically used. This should not be interpreted to mean that other motor types cannot be used for the same type of applications. One motor is called a "brushless DC servomotor," for example, but DC moving coil motors and AC induction motors can also be used as servomotors.

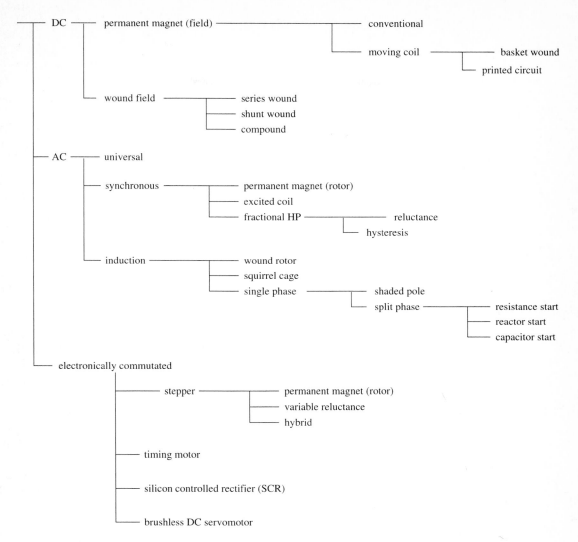

Fig. 3.5 Motor classifications

3.5 CONTROL OF MOTORS

Motors must be started and stopped efficiently. In automated processes, the angular position, speed, and/or acceleration of a motor also must be controlled, despite variations in the loading of the motor. In most of these applications the main controller is a digital computer, which provides control signals to a motor controller, which in turn provides power to the motor.

3.5.1 DC Motors for Control Applications

We have already discussed how DC motors work. We have seen that:

- the armature rotates because of torque caused by current in the armature windings, which are surrounded by the magnetic field, but
- the speed is limited by the generation of Counter Electro-Motive Force as the armature moves in the field.

There are thus two basic methods to control the speed of DC motors:

- controlling the amount of current that flows in the armature, and
- controlling the strength of the magnetic field.

These control techniques will be discussed, but first let's look at the various types of DC motors shown in figure 3.5.

The **permanent magnet** motor can have DC power applied to the armature (via the commutator), but the strength of the permanent magnetic field can't be varied. This type of DC motor, which is the cheapest and the most common in small sizes, can be controlled only by controlling the current in the armature. We will see that this method of control will allow only speed to be decreased from the "nominal" value listed on the motor's nameplate. The direction that this motor rotates can be changed by changing the polarity of the applied DC.

Torque output of a standard permanent magnet motor is limited by the amount of current that the armature windings can carry without burning out. Lots of current carrying capacity means heavier armatures. Figure 3.6 demonstrates the speed versus torque relationship for a permanent magnet motor at a given supply voltage. **Printed circuit** and **moving coil** motors are permanent magnet motors that include no iron in their armatures, thus reducing armature weight (and therefore inertia). Printed circuit motors have "windings" that are little more than circuit board tracks. Moving coil motor armatures consist of woven copper wire windings, set in epoxy to hold their shape. These motors usually rotate very rapidly, and external gearing trades away speed for increased torque.

There are three types of **wound field DC** motors. The first, with speed/torque characteristics as shown in figure 3.6, is the **series wound.** In this type of motor, the field

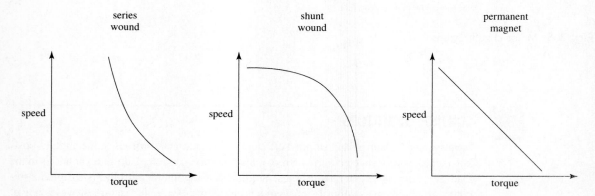

Fig. 3.6 Speed/torque output relationships for DC motors. Voltage is constant in each case.

windings and armature windings are wired in series. Current passing through the field windings must also pass through the commutator to the armature windings. Reducing the DC current to the field also reduces armature current. Since reducing armature current reduces speed, while reducing field strength increases speed, control of this type of motor is difficult. In fact, if the motor is allowed to run without a frictional load, it can accelerate all by itself until it self-destructs. It is also an interesting fact that the direction of rotation of a series wound DC motor *cannot* be changed by changing the polarity of the DC supply. When armature current direction is changed by reversing DC polarity, the magnetic field polarity also reverses, so the induced torque direction remains the same.

Another type of wound field DC motor is the **shunt wound** motor. In the shunt wound motor, the field windings and the armature windings are brought out of the motor casing separately, and the user connects them to separate supplies so that the field strength and the armature current can be controlled independently. This type of motor can, depending on the type of control selected, have its speed reduced or increased from the nominal values. The direction of rotation of this type of motor can be changed by changing the polarity of either, but not both, of the supplies. The speed/torque relationship for this type of motor at a given voltage supply is shown in figure 3.6.

It is possible to tap into a series wound DC motor's circuitry so that shunts can be added to allow some of the DC power to bypass either the field windings or the armature. This yields a **compound wound** motor. Speed can be controlled by varying resistance in the shunts. Speed control is limited, since reduction in the resistance in a shunt (e.g., shunt at armature) means increased current in other parts of the circuit (e.g., through the field windings). Overheating is then a danger. This method of controlling compound motors is used in applications where two or more motors must have their speeds synchronized, and only slight variation of motor speed is required.

Speed Control for DC Motors

Two simple formulas are important in understanding the response of DC motors to changes to supplied power. The first,

$$RPM = \frac{V_a - (I_a * R_a)}{F}$$

where:

RPM = Motor rotational speed
V_a = Voltage across the armature
I_a = Current in the the armature
R_a = Resistance in the armature
F = Field strength

shows the relationship of the motor speed to applied armature voltage (decreased V_a will decrease RPM), to armature resistance (adding resistance to R_a will reduce RPM), and to the field strength (a decrease in F will *increase* speed). Armature current is slightly harder to control because of the variation in CEMF as speed increases, but control of speed through current control is often the method of choice. It allows maximizing torque without exceeding the maximum current range of the motor.

The second equation,

$$T = K * F * Ia$$

where:

T = motor output torque
K = a constant for the motor

shows the relationship between torque, field strength, and armature current. Note that a decrease in field strength, which we said earlier would cause an increase in speed, will also cause a reduction in the available torque unless armature current is increased. In fact, reducing field strength reduces CEMF, so armature current does increase.

These two equations can be used in control of DC motors in several ways. A motor's name plate will include a nominal speed. **Armature voltage control** can be used to reduce the motor's speed from this nominal value. Controlling the voltage across the armature controls the current. Since armatures are made as light as possible, it is practical to use this type of control only to reduce the current from nominal; otherwise the light armature windings might overheat and burn out. This type of motor control is thus used to reduce the speed of the motor from its nominal value. Figure 3.7 shows how armature voltage control can be used in a servosystem controlling motor speed. In this servosystem, sensors monitor motor speed. Any difference between the DC speed setpoint and the DC feedback (which is proportional to actual speed) will cause the armature voltage supply to change to correct that error.

The other method of DC motor speed control, **shunt field control,** is used to increase the motor's speed above the nominal speed. Voltage applied to the field windings is *reduced* from the nominal maximum. The reduction in electro-magnetic field strength leads to a reduced amount of CEMF generated in the armature windings. With lower CEMF, more current flows in the armature windings, so more torque results and the motor accelerates. As the motor speed increases, CEMF increases and torque drops until torque is again just sufficient to move the load. At this new speed, more current is moving in the

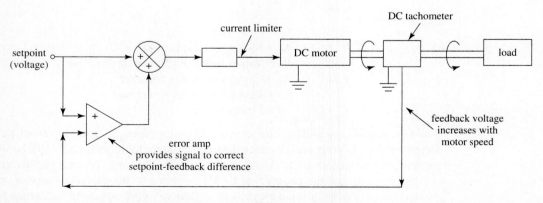

Fig. 3.7 A motor with its speed controlled by a servosystem

armature than before the field voltage was reduced. Shunt field control is typically used to increase speed to no more than 400% of nominal speed. Beyond this point, increased armature current can damage most motors.

Stopping of DC Motors

Stopping of a motor is a form of speed control. Methods used to stop DC motors are similar to speed control techniques. To stop a motor, you must accelerate it in the direction opposite to that in which it is moving. Mechanical brakes can be used, but will not be discussed here.

There are two methods of electrically stopping a DC motor. The most common is the **dynamic braking** method. To stop a DC motor using this method, the magnetic field remains in place, but the armature voltage supply is replaced by a resistor. The motor thus becomes a generator, and its kinetic energy is converted to electric current that burns off in the resistor. Small motors can be stopped in milliseconds using this technique.

The other electrical method to stop a DC motor is called **plugging.** As in dynamic braking, the magnetic field must be retained until the motor comes to a stop. Unlike dynamic braking, the armature is connected to a DC supply of opposite polarity. This results in a dramatic acceleration in the opposite direction. Armature current due to the (reversed) armature DC supply is further increased by the current due to the existing CEMF! Braking is rapid, but the high current is hard on the armature. This type of braking is therefore used only for emergency braking or for the braking of motors specially built for this type of extra armature current.

Plugging carries another potential problem. Once plugging stops a motor, it will then accelerate the motor in the opposite direction. Motors that are stopped by plugging need to have a **zero speed switch** mounted on the motor shaft or on the load. A zero speed switch contains an inertial switch that disconnects the armature voltage supply when motor speed reduces to a near stop.

These electrical braking techniques will not *hold* a motor in the stopped position, so if positive braking is required to hold a load steady at the stopped position, then:

- A mechanical brake can be used to very rigidly hold a load wherever it happens to stop, or
- A positional servosystem can cause the motor to stop at a given position.

With many servosystems, a certain amount of play (or "backlash" or "compliance") must be tolerated. The amount of play will depend on the control system. Electrical robots have positional servosystems, and some can be held at a given position to within less than a thousandth of a degree. Chapter 6 describes servosystems.

Methods of Controlling DC Power Levels

There are two basic methods of reducing DC power to control motor speed. (Other techniques can convert AC to controlled DC voltage levels. See the section on electronically commutated motors later in this chapter.) The first method is a wasteful method that is to be avoided. A portion of the supplied DC can be *burnt off with resistors,* leaving the desired DC level. If DC is regulated by **operational amplifiers** instead of by variable resistors, the result is the same: power is wasted in the op amp.

The electronic revolution has given us a less wasteful method. This second method is to use a "switched" power supply, and is sometimes called **Pulse Width Modulation (PWM).** The basic principle of PWM is shown in figure 3.8. In PWM, the load is rapidly connected and disconnected from the supply. There is never any need to burn off part of the supplied power. If full power is applied to the motor only half the time, the motor runs at the speed at which it would run if half power was applied. The widths of the "on" pulses can be varied, or "modulated," so that the average power to the motor is anywhere from fully off to fully on.

Digital computers are quick enough that they can periodically output a signal to actuate or deactivate a power switch, without seriously interfering with other programs they may be running. There are also individual integrated circuit chips available that can control the speed of a motor relative to a setpoint supplied by the digital or analog controller. Figure 3.9 shows how one of these devices could be used in a motor control servosystem. The chip uses as inputs: the DC setpoint for the desired speed, and the DC voltage output from the speed sensor. Inside the chip, these two inputs are subtracted, and the resultant "error" voltage dictates how wide the output pulses will be. The low voltage pulses that this chip outputs are used to switch the larger power source for the motor.

3.5.2 AC Motors for Control Applications

Alternating current is current that is driven forward then backward by an alternating voltage. In one cycle the AC voltage rises from zero volts to a peak, returns to zero, continues to a negative peak, and returns to zero. It is possible to control the amplitude of the AC voltage peaks and to control the frequency of the AC cycles.

AC in North America is supplied at 60 hertz (60 cycles per second), at a variety of nominal voltages (120V, 240V, 580V, and higher for special applications), as single phase or three phase. The actual voltage received can and does vary with distance from the source. Two phase AC, used for some motors, must be "manufactured" in the motor controller used with the motor. Figure 3.10 shows what this terminology means.

AC motors can sometimes be controlled by adjusting the amplitude of the AC power supply, but the most common control method is to vary the frequency of the AC supplied to the motor.

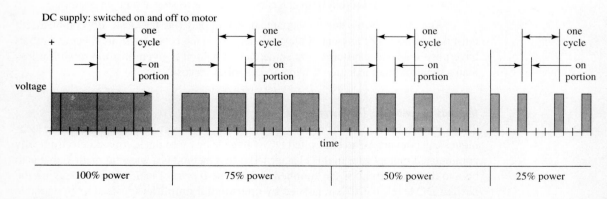

Fig. 3.8 Simple pulse width modulation

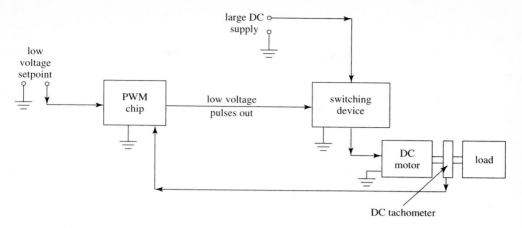

Fig. 3.9 Pulse width modulation chip

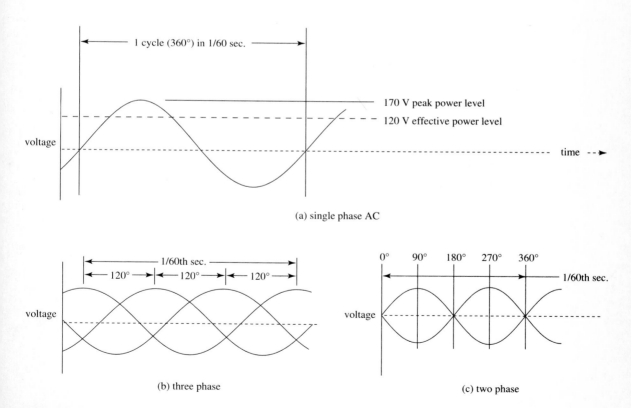

Fig. 3.10 AC waveforms: (a) single phase; (b) three phase; (c) two phase.

3.5 Control of Motors

In figure 3.5, we divided AC motors into three types:

- universal
- synchronous
- induction

As engineers become more comfortable with our new ability to control AC frequency, AC motors are becoming more commonly used in control applications.

The most often used AC motor, when control of torque, speed, or position is required, is the **induction** motor. This motor is called an "induction" motor because current is induced into the rotor's conductors by a magnetic field that moves past them. This current, which moves through the magnetic field, causes torque, so the rotor turns. Let's look at this more closely.

The induction motor's magnetic field is caused by AC power in the field windings. The magnetic field rotates around the rotor. Figure 3.11 shows how three phase AC supplied to the field windings of an AC motor causes the field to rotate. Since the field is initially rotating around a stationary rotor, then the rotor windings are moving relative to the field. Current is induced into the conductors (remember the right hand rule).

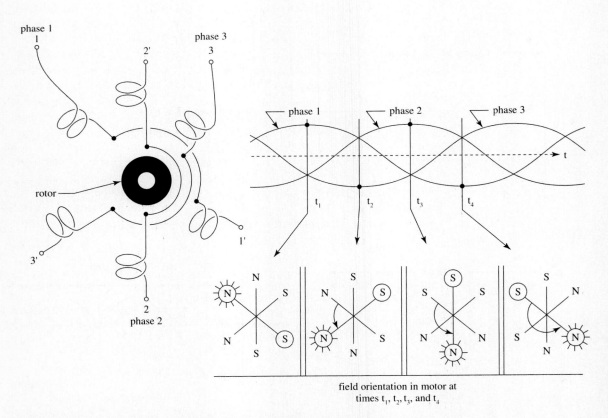

Fig. 3.11 Rotation of the field in an AC motor supplied with three phase AC

The conductors may be windings of copper wire in a **wound field** induction motor. They may be copper bars, parallel to each other and connected to a common copper ring at both ends. This copper bar and connecting ring arrangement, when seen without the iron core of which it is part, looks like the exercise ring used by pet mice or (presumably) pet squirrels. Hence the name for this type of motor: the **squirrel cage** motor.

Note that current is induced into the rotor and no commutator or slip rings are necessary. This motor is therefore brushless. If DC power is electronically switched around the field coils instead of having an AC supply drive the motor, then the motor could be called a brushless DC motor.

The more the rotational speed mismatch between the moving magnetic field and the rotor, the more current is induced in the rotor. When the rotor is stationary, therefore, there is a large current induced. This current, moving in the windings or squirrel cage, moves across the lines of magnetic flux. The left hand rule tells us that current moving in a field will result in a force on the conductor, and therefore torque to accelerate the rotor. The more current, the more torque. The torque will accelerate the rotor in the same direction that the field rotates.

As the induction motor's rotor accelerates, the speed mismatch between the field and the rotor reduces, so the induced current reduces, and therefore the torque to accelerate the rotor reduces. Eventually, the rotor will be rotating almost as fast as the field, and there will be only enough torque to keep the rotor and the load moving, but not enough to cause more acceleration. The final speed mismatch between the field and the rotor is called *slip* in an induction motor. Slip can never be 0%, or torque also would be zero, and the rotor would decelerate. The amount of slip increases as load increases, because more torque is required, and torque can be generated only if the rotor moves slower relative to the field. Typical slip in an induction motor is between 1% and 5%. Figure 3.12 shows the speed versus torque relationship for an AC induction motor. Note that torque peaks a little below the speed at which the field rotates. The dotted line shows that for short periods, such as at starting of the motor, high current and resultant high torques can be allowed. The inclusion of a thermal sensing switch in the circuit will ensure that the motor does not continue to generate damaging current in the rotor for extended periods.

In figure 3.11, we saw that three phase AC, if each phase is wired to a separate set of poles in an AC motor's field, will cause a rotating field. If we tried the same type of simple connections using single phase AC, figure 3.13 shows that the field simply

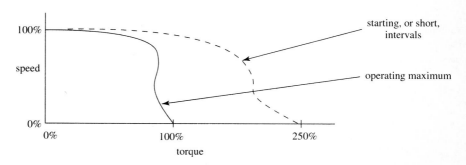

Fig. 3.12 Speed/torque relationship for an AC induction motor

alternates; it does not rotate. We need some way of defining the direction of the rotation, so that a **single phase** AC motor can operate.

The **shaded pole,** single phase, AC induction motor uses a cheap but energy wasteful method to define the direction of rotation of the field. This type of construction is common in small motors where energy efficiency is somewhat unimportant. Figure 3.14 shows a copper ring around the smaller of two sections of the same pole in a single phase induction motor. When the AC tries to change the polarity of the whole upper pole to a north, the change in the field induces a current in the copper ring. Current in the ring opposes the setting up of the magnetic field, effectively delaying the rate at which *this portion* of the pole changes to a north pole. The result is that the change in magnetic polarity sweeps across the face of this single pole, defining a direction of rotation for the field. This method is energy-wasteful because it continues to spend energy defining the direction of rotation after the rotor has started. The direction of rotation of a shaded pole motor cannot be changed.

Another type of AC induction motor that can be used with single phase AC power is the **split phase** motor. In this motor, each motor pole has another starting-up pole wired in parallel. The parallel circuits usually have an inertial switch in series with the starting pole. Figure 3.15 shows some motors of this type. In all three cases, the purpose of the inertial switch is to disconnect the "start" winding once the motor has started, to reduce energy waste. Before the start windings disconnect, however, the characteristics of the circuits are such that the magnetic polarity of the start windings either leads or lags the changing polarity of the "run" windings. If the wires for the start winding and the wires for the run winding are brought out of a split-phase motor separately, then the direction of rotation can be changed by reversing the connections of *either* (not both) windings to the AC supply.

In figure 3.15 (a), the start winding of the **resistance start,** split phase, AC induction motor has fewer windings than the run winding. The start winding therefore has less inductance, so the magnetic polarity of this pole changes earlier than the run winding. The direction of rotation of the field is thus from the start to the run windings.

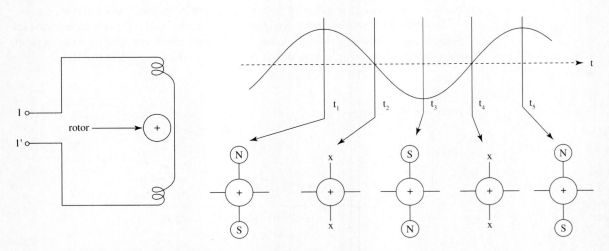

Fig. 3.13 Single phase AC showing alternating magnetic field

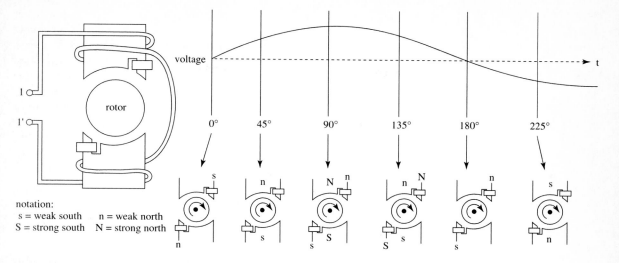

Fig. 3.14 A shaded pole motor showing the direction of rotation of the magnetic field

Figure 3.15 (b) shows a **reactor start** motor. The added inductor in the run winding circuit causes the change in magnetic polarity of the run winding to lag the polarity change of the start winding. The effective direction of rotation of the field is from the start to the run winding.

In Figure 3.15 (c), the **capacitor start** motor has a capacitor in series with the start winding. This combination causes the magnetic polarity of the start winding to lead the run winding. The effective direction of rotation of the field is thus from the start winding

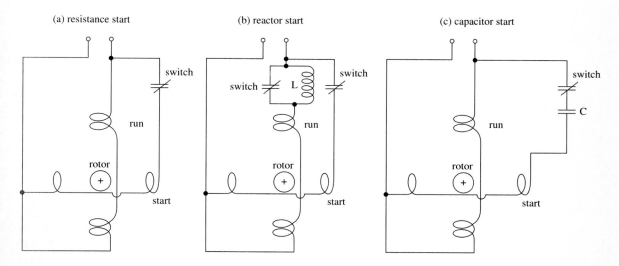

Fig. 3.15 Split phase AC induction motors: (a) resistance start motor; (b) reactor start motor; (c) capacitor start motor.

to the run winding. It should be noted here that in some motors of this type, the capacitor and start winding are not disconnected even after the motor reaches speed. These motors are called **capacitor run** motors, or sometimes **permanent capacitor** motors.

The **synchronous** AC motor is called synchronous because the rotor follows the magnetic field *exactly* as it rotates around in the housing. Figure 3.11 showed how three phase AC supplied to the field windings causes the magnetic field to rotate. The synchronous motor's rotor will "lock onto" the magnetic poles of the field because the rotor has magnets as well. The magnets in the rotor may be **permanent magnets.**

Note that the field we examined in figure 3.11 inherently rotated at 3600 RPM (with 60 Hz AC). There are three ways a stationary permanent magnet rotor can be accelerated to the speed at which it can catch and lock onto the rotating field:

- Some synchronous motors require small auxiliary motors to accelerate them to the speed at which they can "lock on."
- Others are caused to accelerate by the inclusion of squirrel cages in their rotors so that they start as if they were induction motors.
- With control of the frequency of the AC supply, it is possible to control the speed of field rotation so that it accelerates slowly enough that the rotor can stay locked on.

Fractional hp synchronous motors can be made without magnets in the rotor. There are two types of these low-torque motors: hysteresis and reluctance.

The **hysteresis** motor has a rotor of cobalt steel that becomes magnetized in the presence of another magnetic field, but which has a high magnetic hysteresis (does not change its magnetic polarity easily). Once these rotors have been accelerated to the speed at which they can lock onto the rotating field, they will follow the rotating magnetic fields as if they had permanent magnets in their rotors.

Reluctance motors have high spots on their iron rotors, and these high spots tend to align in the direction of the magnetic field. As the magnetic field rotates, so does the rotor. Although this type of motor is classified as a fractional hp motor because its torque at synchronous speed is low, some are available with squirrel cages so that they can produce as much as 100 hp at lower speed.

Excited coil synchronous motors have windings in their rotors that can be provided with DC so that the rotors then have electro-magnetic poles. DC is supplied to these windings only after the rotor has been accelerated to locking-on speed. Below locking-on speed, the unpowered windings act as a squirrel cage to accelerate the rotor. These motors need special speed-sensing circuitry to ensure that the DC is turned on to the rotor at the right speed and the right time.

The **universal** motor is similar in construction to a series wound DC motor. It will run on either DC or AC power. The universal motor is the only AC motor with a commutator, through which the armature is connected in series to the field windings. When AC power is supplied, the magnetic polarity of the field windings changes as the direction of the alternating current changes. When the magnetic field polarity changes, the direction of current flow in the armature also changes, so that torque is still generated in a consistent direction. As with the series wound DC motor, this motor will always run in the same direction (unless the wiring from the field windings to the armature is reversed).

This type of motor cannot be easily controlled, either by controlling AC amplitude or frequency but their speed can be controlled if the amplitude of the AC supply is adjusted. Portable handtools with variable speed controls are usually universal AC motors.

Control for AC Motors

There are now several inexpensive and successful techniques for varying the frequency of AC power. This was not always the case. Prior to the development of silicon devices, AC motor speed was dictated by the 60 Hz supply frequency.

One early way that AC motors had their speed changed was by *changing the number of field poles*. This method is shown in figure 3.16. Figure 3.16 shows the location of a north pole as the field rotates. The rotation rate of a motor can be doubled by changing how the poles are connected to the three phase supply. In fact, the rate of rotation of the field for any AC motor can be calculated using the following equation:

$$N = \frac{120 f}{P}$$

where:

N = speed of rotation of the field in RPM
f = frequency of the AC in Hz
P = number of separately-connected sets of poles

e.g., with an AC frequency of 60 hertz. There are two complete sets of poles in figure 3.15 (a), so the speed of rotation of the field is:

$$N = (120 * 60)/2 = 3600/2 = 1800 \text{ RPM}$$

There is only one complete set of poles in figure 3.15 (b), so the field rotates at:

$$N = (120 * 60)/1 = 3600 \text{ RPM}$$

It is important to note that this method of controlling the speed of AC motors will not allow true variable speed control. The speeds that can be obtained are tied to the 60 Hz supply and limited by the small number of poles that can be physically fitted into a motor's casing.

With the development of cheap silicon-based components such as diodes, and triggered conductive devices such as silicon controlled rectifiers, **control of the frequency** of AC power has become possible.

There are three methods of controlling the frequency of AC supplied to an AC motor. Two of the methods require a DC supply, usually provided by a separate circuit made up of silicon devices that "rectify" an AC supply. The third method allows the AC supply to be converted directly to AC with a new, lower frequency.

One of these three methods is a variation of **Pulse Width Modulation (PWM)**. Remember that when we discussed speed control for DC motors, we examined PWM for control of DC level. The DC was switched on and off, and the output DC level was equal to the average of the output DC over a short time.

PWM for control of AC frequency is very similar, except that instead of using the "on" and "off" states, PWM for AC uses (at least) the "on-positive" and the "on-negative" states. Varying the proportional length of the positive and negative pulses causes

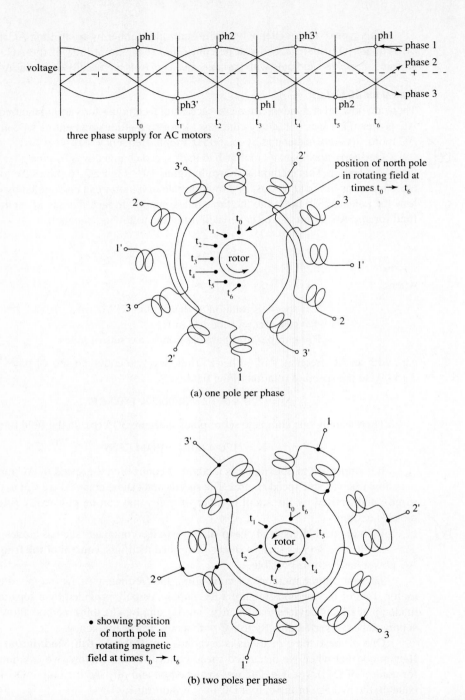

Fig. 3.16 Three phase supply for AC motors, and changing the speed of a six pole AC motor by changing connections to the poles: (a) Each pole connected separately to each phase; (b) poles connected in pairs to each phase. (Location of the north pole is shown in figures at times t0, t1, t3, t4, t5, and t6.)

the averaged output voltage level to rise and fall. It is a simple matter to program a microprocessor to control DC in this way, producing controlled AC frequencies.

For finer control of the output AC, a third state (off) can be included in the PWM cycle. For even finer control, two different positive levels and two negative levels can be used with the "off" level, as shown in figure 3.17. The top part of figure 3.17 shows the output from the PWM circuitry, and the bottom portion shows the output voltage level after being averaged in a resistor-capacitor filter circuit. The four DC levels, plus the off state, can all be obtained with electronic switching of components in a common circuit for rectification of AC. Getting more than these four DC levels from a single AC rectifier is more difficult.

The second basic method of inverting DC so that the result is approximately AC has also sometimes been called PWM because the principles of PWM are used again. It is more properly called **variable voltage inversion** or sometimes the **six step method.** As shown in figure 3.18, four different DC levels are available (two positive and two negative). The four different DC levels are switched to the motor in a sequence of six steps, so that a sinusoidal waveform is approximated. The width of each step is controlled, so that AC frequency is controlled.

The **cycloconverter** method of frequency control requires AC input that, as shown in figure 3.19, is adjusted in three steps.

1. The incoming AC is rectified so that several cycles become positive and then several become negative.

2. The waveforms are clipped. (In fact, the circuitry that rectifies the incoming AC also does the clipping.) The earliest and latest pulses of each polarity are clipped more severely than the middle pulses.

3. The rectified and clipped waveform is then filtered. The result is an AC waveform of a frequency less than the incoming AC.

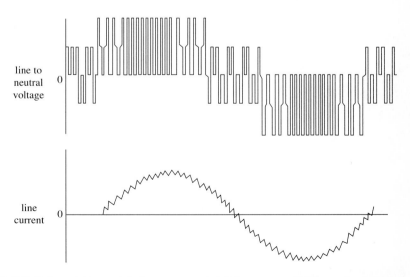

Fig. 3.17 Pulse width modulation to control AC frequency. (Diagram by permission, Allen-Bradley Canada Ltd., A Rockwell International Company.)

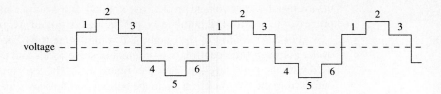

Fig. 3.18 Six step variable voltage inversion to control AC frequency

Stopping of AC Motors

AC induction motors can be brought to a rapid halt by **dynamic braking**. The AC supply to the field is replaced with a DC supply. The motor becomes a generator, but the current generated in the rotor has no route to the outside world, so it must flow around the short-circuited squirrel cage or wound field, converting the rotational energy into heat.

Synchronous motors without induction coils or squirrel cages in their rotors cannot be stopped dynamically. Mechanical brakes are necessary.

3.5.3 Electronically Commutated Motors

Figure 3.5 listed four types of electronically commutated motors: the stepper motor, the timing motor, the SCR motor, and the brushless DC servomotor. These motors have

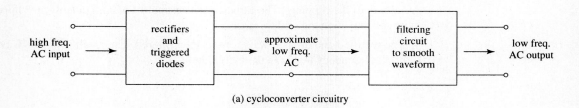

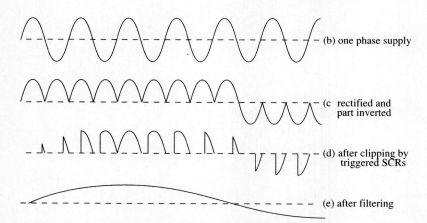

Fig. 3.19 The cycloconverter method of producing frequency controlled AC

built-in electronic control and/or require electronic motor controllers. Many motors not included in this classification can also be controlled electronically.

The term *electronic commutation* implies that electronic circuitry replaces a commutator. "Commutator" suggests a DC motor, which is why electronically commutated motors are often classified as DC motors. We will use "electronic commutation" to describe any electronic switching of circuitry as required to drive a motor. An electronically commutated motor, therefore, doesn't necessarily have to resemble the DC motors studied so far.

Stepper Motors

In terms of construction, **stepper motors** have much in common with AC synchronous motors. They come in three types:

- The **permanent magnet** rotor type, as you might guess, has a rotor with permanent magnets that follow the magnetic field as it rotates around in the field windings. An advantage of this type is that they have a "detent torque," or a tendency to remain magnetically locked in position even when power to the motor is off.
- The **variable reluctance** type has a toothed rotor, so that the high spots on the rotor tend to follow the rotating field.
- The **hybrid** type has both permanent magnets and a toothed rotor.

The electromagnetic poles in a stepper motor's housing are designed such that they can be individually switched to DC. The rotor is supposed to "step" from one activated pole to the next as the poles sequentially turn on and off. A separately-supplied stepper motor controller controls in which order the windings receive DC. Stepper motor controllers "ramp" motors from one speed to the next by gradually changing the rate at which steps are supplied. They also allow motors to be operated in "slewing" mode, during which the rotor rotates without stopping after each step.

Basic stepper motors have an inherent step angle, or angle between possible step positions. Several techniques are now used to energize combinations of poles, and to control the DC level supplied to individual poles, so that the effective location of the magnetic field is between poles. As a result, "micro-steps" of fractions of a degree can be obtained.

Stepper motors are most commonly used in "open loop" position control. No position sensors are used, but as long as the rotor does not miss any steps (i.e., available motor torque is adequate), the position of the rotor is known. Stepper motors have also been used in systems with position sensors.

Timing Motors

Electronically commutated **timing** motors were one of the first applications of built-in electronic commutation. They contain internal controllers to ensure they always run at the same speed, and are thus self-contained servosystems.

These motors are similar to AC synchronous motors. They have permanent magnetic rotors and a field that rotates, causing the rotor to follow at the same speed. Timing motors differ from AC synchronous motors in that the current is switched from field pole to field pole ahead of the rotor by silicon-based components on a circuit board built into the motor.

Figure 3.20 shows that the position of the rotor must be detected so that the current can be switched to the next pole at the right time. Hall effect switches, built into the motor housing, are often used to detect the rotor position. When the permanent magnetic rotor reaches a Hall effect sensor, the output from the sensor switches DC current to the next pole. If the rotor is delayed, the controller increases the amount of DC supplied to the next pole, so that motor speed will increase. DC is reduced if the motor runs too fast.

SCR Motors

SCR motors are DC motors, but operate when supplied with AC current! Built-in circuitry, containing silicon controlled rectifiers (SCRs), "rectify" the AC supply to yield effective DC to operate the motor.

The SCRs also block part of the AC waveform. Figure 3.21 shows the input waveform as well as the rectified and modified waveforms. The average voltage level reduces from "full DC" down to zero as more of the waveform is blocked. A variable resistor setting, or a small DC control voltage, determines how much of the waveform is blocked.

An alternative type of SCR motor control circuit does not remove any of the waveform, but only inverts the input waveform at a variable time during the AC cycle. The average DC varies, as shown in figure 3.22, from full positive DC to full *negative* DC. The motor can thus be reversed.

SCR motor control circuits usually contain a feedback loop detecting the CEMF being generated, and use this feedback to modify the time at which the SCRs trigger, like the electronically commutated timing motor. If the motor is going too fast, producing more than the correct amount of CEMF, the SCRs delay firing, blocking more of the waveform, and the motor slows down. If the motor is going too slow, the SCRs fire earlier, passing more current to accelerate the motor.

Brushless DC Servomotors

Brushless DC servomotors are like SCR motors in that they are really DC motors (permanent magnet DC motors), and in that their controllers convert AC to DC and modify the effective DC level in response to a command signal. Built-in position sensing allows

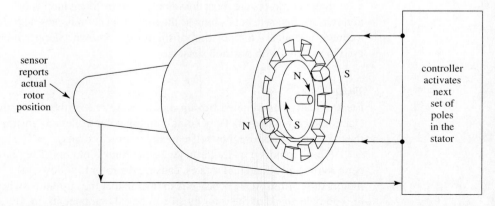

Fig. 3.20 DC control and switching in an electronically commutated timing motor

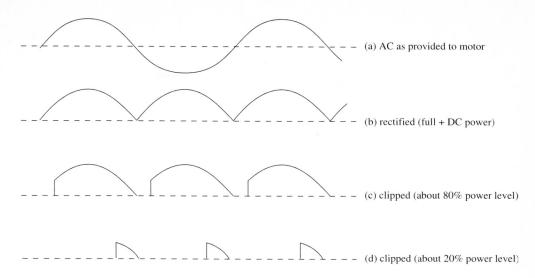

Fig. 3.21 SCR control of AC to produce controlled DC levels

closed-loop control of speed. Brushless DC servomotors differ from standard PM DC motors in that they are constructed to maximize torque without increasing rotor inertia.

A motor used in a speed or position control servosystem requires a speed or position sensor. Like most other servomotors, a brushless DC servomotor has a built-in rotor position sensor.

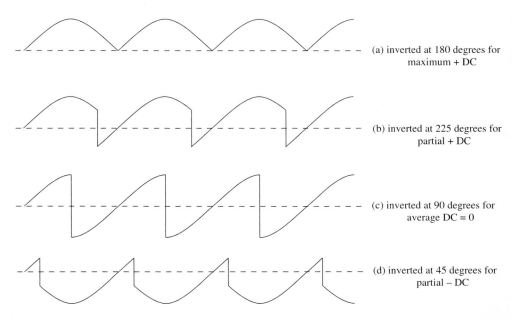

Fig. 3.22 SCR inversion of AC for effective DC control

3.5 Control of Motors

A servomotor should have a high torque-to-inertia ratio, so that it can accelerate quickly. A standard permanent magnet DC motor can be built for greater torque by increasing the amount of conductor in the armature so that more current can be moved across the magnetic field. Unfortunately, this also increases the inertia of the armature, so more torque is required to accelerate the rotor!

The previously-discussed printed circuit and basket wound DC motors are often used as servomotors. Their armatures contain nothing but conductors, yet these motors still have an unsatisfactory torque/inertia ratio for many applications. Robots built with these types of motors usually also have high ratio gear drives (often harmonic drives, see chapter 2) to trade away speed to gain torque.

There are two brushless DC servomotor solutions to the torque/inertia dilemma. Both solutions also eliminate the commutator and brush.

In the type of brushless DC servomotor shown in figure 3.23 the *armature is held stationary and the permanent magnetic housing is allowed to rotate about it*! Since the inertia in the armature is now unimportant to the motor's acceleration, it can be built heavy to carry very high currents, so tremendous torques can be generated. Electronic commutation to switch the DC from one stationary winding to the next makes this construction possible. Hall effect sensors detect the position of the magnets in the rotating field and ensure DC is switched to the correct stationary armature winding. Recent innovations in the production of lightweight magnets, used in the rotating housing, have helped as well. These new brushless DC motors, used without gearing, are finding applications in many modern **direct drive** servosystems.

A further improvement in the design of brushless DC servomotors, shown in figure 3.24, has the DC *motor turned inside out*. The armature windings have been moved into the stationary housing, surrounding the rotor. The permanent magnets are mounted in the

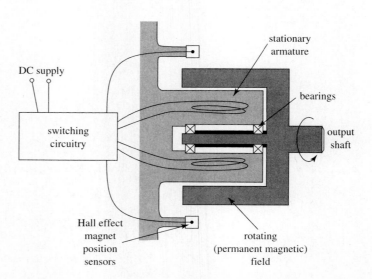

Fig. 3.23 Brushless DC servomotor with stationary armature

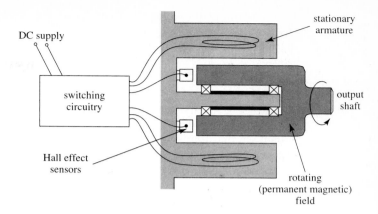

Fig. 3.24 Brushless DC servomotor turned inside out

rotor. Since this rotor can have a smaller diameter than that in figure 3.23, it has even less rotational inertia.

3.6 SUMMARY

We have looked at the **history** of electric motors, at a **classification system**, and at the motors and control techniques themselves. The classification system is limited to the most common motor types used in automated systems today.

We looked at **DC motors**, of which perhaps the most common type in use today is the **permanent magnet** type (which includes brushless DC motors). These motors can be controlled using **armature voltage control**, which can be used only to reduce motor speed to below nominal speed. **Wound field** motors are used if we also require **Field Voltage Control** in order to increase motor speed from the nominal value. Voltage can be controlled in either case by burning off some of the supply DC voltage, or by using a switched power supply. Modern DC motor control circuits use silicon-based devices to rectify AC and to then switch it on and off to control the effective averaged DC levels. This method is known as pulse width modulation (PWM).

In the sections on **AC motors**, we saw that field windings supplied with multi-phase AC result in rotating magnetic fields. We saw how rotors can be made to rotate even if single phase AC is used. We examined the very popular **induction** motor, which includes squirrel cage motors and the less common wound rotor types. We saw why rotors in induction motors rotate with a slip of between 1% and 5%. **Synchronous** motors, on the other hand, have rotors that follow the rotating field exactly. The preferred method of controlling the speed of both induction and synchronous motors is by **controlling the frequency of the AC** supplied to the motor. The three basic methods of controlling AC frequency are **PWM, variable voltage inversion** (the six step method) and, less commonly, the **cycloconverter** method. This last method is limited in that it can output only new AC frequencies well below the input AC frequency.

Electronically commutated motors are new versions of the motors already discussed. **Stepper** motors were originally intended for open loop operation. Stepper motor controllers sequence DC around stepper motor field windings, and the rotor "steps" with the magnetic field. The **timing motor** has a built-in circuit board to switch the DC supply around the field windings, and otherwise operates like a synchronous AC motor. The **SCR motor** is a DC motor, with the DC supplied by an SCR-based circuit that produces variable DC by chopping and/or inverting AC.

Servomotors differ slightly from other motors. They may have built-in speed or position sensors. They often have light rotors for rapid acceleration, and often have low torque output so they require significant gearing to be useful. Brushless DC servomotors are increasingly used because they have high torque output even at low rotor speeds.

REVIEW QUESTIONS

1. How does a rotor differ from an armature?
2. How do slip rings differ from a commutator and brushes?
3. Why does the armature of a standard DC motor, with a permanent magnetic field, rotate?
4. Why does the rotor of an AC induction motor rotate?
5. How would you cause a DC motor to rotate faster than its nominal rotation speed? Identify one type of DC motor in which this is impossible.
6. Why can a (true) brushless DC servomotor motor be used in direct drive applications requiring high torque and high acceleration, while standard DC motors can't?
7. Describe two methods of electrically stopping a DC motor. Which method requires a zero speed switch and why?
8. How can pulse width modulation (PWM) be used to output 8 volts DC from a 24 VDC supply? Use a sketch of the PWM output.
9. If 120 volts AC at 60 Hz can cause a synchronous motor to rotate at 1800 RPM, what would you change in the AC signal to increase the motor speed?
10. Describe variable voltage inversion as a technique for motor speed control.
11. If you must start a synchronous motor by suddenly connecting it to the 120 volt, 60 Hz AC supply, what extra feature do you require in the motor?
12. Identify one extra feature that a single phase AC motor must have that a three phase AC motor doesn't need. Why is this feature needed? Several variations were described in this chapter. Pick one of the variations and explain how it allows the motor to be accelerated from a stop condition.
13. What causes a stepper motor to rotate?
14. Describe how an SCR motor uses its AC supply to control speed.
15. Describe two types of brushless DC servomotors. Identify two advantages of brushless DC servomotors versus standard DC motors.

CHAPTER 4

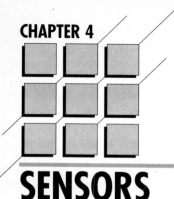

SENSORS

CHAPTER OBJECTIVES

After completing this chapter, the reader will be able to:

- Understand the meaning of sensor descriptors including range and span, resolution, repeatability, and linearity.
- Discuss the difference between switches, switch arrays, transducers, and switched transducers.
- Describe the operating principles of a wide range of sensors including:
    - proximity sensors (inductive, capacitive, and optical)
    - temperature transducers (RTD, thermistor, IC, and thermocouple)
    - differential versus gauge pressure sensors
    - strain gauges
    - several flow sensors
    - potentiometers
    - linear variable differential transformers
    - magnetostrictive sensors
    - waveform-type position sensors (optical, ultrasound, and laser interferometer)
    - resolvers and synchros
    - incremental and absolute encoders
    - position sensors used as velocity or acceleration sensors
    - velocity sensors (AC generator, tachometer, and Doppler type)
    - accelerometers

▬▬ Discuss some of the complex sensing systems that the above sensors can be integrated into, including systems for:
- automated parts gauging and coordinate measuring machines
- bar code readers

▬▬ Recognize the potential for reading errors where there are rapid repeated changes in the measured conditions.

4.0 SENSORS

Good sensors are essential in any automated system. Computer Integrated Manufacturing (CIM) is possible only if computerized controllers can detect the conditions in manufacturing processes from beginning to end. Some sensors detect only part presence. Other sensors, bar code readers, for example, help to track materials, tooling, and products as they enter, go through, and leave CIM environments. During manufacturing, automated gauging of machined parts as they exit Direct Numerical Controlled (DNC) machining centers allows Statistical Process Control (SPC) of unattended manufacturing processes. In fact, *every automated manufacturing operation should include a sensor* to ensure it is working correctly.

4.1 QUALITY OF SENSORS

The best sensor for any job is one that has sufficient quality for the job, has adequate durability, yet isn't any more expensive than the job requires. **Spec sheets** (short for "specification sheets") use many terms to describe how good sensors are. The following section discusses what these terms mean.

4.1.1 Range and Span

The range, or "span," of a sensor describes the *limits of the measured variable that a sensor is capable of sensing*. A temperature transducer must have an output (e.g., electric current) that is proportional to temperature. There are upper and lower limits on the temperature range at which this relationship is reasonably proportional. Figure 4.1 shows the concept of span in describing the usefulness of a resistance-type temperature sensor. Current flow through the sensor varies with temperature between t_1 and t_4, but is only reasonably proportional to temperatures between t_2 and t_3. To be able to claim reasonable proportionality, the sensor's supplier would actually quote the more restrictive temperature span of from t_2 to t_3.

4.1.2 Error

The error between the ideal and the actual output of a sensor can be due to many sources. The types of error can be described as resolution error, linearity error, and repeatability error.

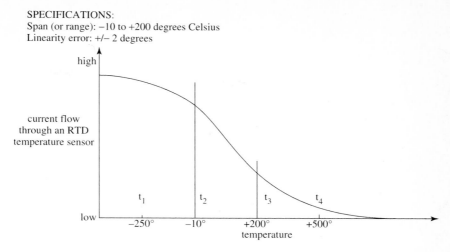

Fig. 4.1 Range and linearity error in a temperature sensor

Resolution

The resolution value quoted for a sensor is *the largest change in measured value that will not result in a change in the sensor's output.* Put more simply, the measured value can change by the amount quoted as resolution, without the sensor's output changing. There are several reasons why resolution error may occur. Figure 4.2 demonstrates resolution error due to hysteresis. Small changes in measured value are insufficient to cause a change in the output of the analog sensor. Figure 4.3 demonstrates resolution error in a sensor with digital output. The digital sensor can output 256 different values for temperature. The sensor's total span must be divided into 256 temperature ranges. Temperature changes within one of these subranges cannot be detected. Sensors with more range often have less resolution.

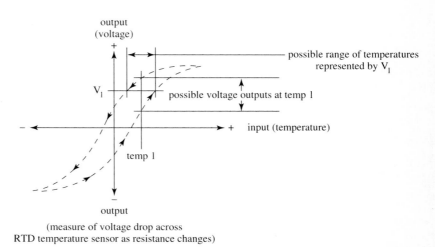

Fig. 4.2 Hysteresis and resolution in a temperature sensor

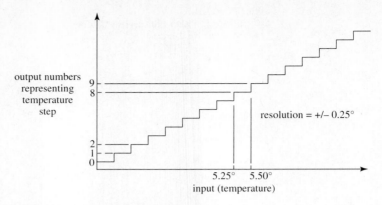

Fig. 4.3 Resolution in a digital temperature sensor

Repeatability

Values quote for a sensor's repeatability indicate *the range of output values that the user can expect when the sensor measures the same input values several times.* The temperature sensor in figure 4.2, for example, might output any value from a voltage value slightly higher than V_1 to a value slightly lower when the actual temperature is at temp 1. Repeatability does *not* necessarily mean that the output value accurately represents the sensed condition! Looseness in mechanical linkages are one source of repeatability error.

Digital sensors can have very impressive repeatability values, even if they have poor resolution. While the sensed temperature of the sensor whose output is shown in figure 4.3 is between the limits of a single step of resolution, it is always reported as being at that same step.

Linearity

The ideal transducer is one that has an output exactly proportional to the variable it measures (within the sensor's quoted range). No transducer's output is perfectly proportional to its input, however, so the sensor user must be aware of *the extent of the failure to be linear.* Linearity is often quoted on spec sheets as a +/− value for the sensor's output signal. Figure 4.4 demonstrates what a linearity specification of +/− 0.5 volts for a pressure sensor means. Note that the quoted value for non-linearity represents the worst case, though the sensor's linearity error is not actually that bad over its whole range. Linearity can often be improved by using only the central portion of the sensor's range, as linearity error often is worst at the extremes of that range. Digital sensors inherently have a degree of non-linearity, as can be seen in the shape of the line in figure 4.3.

4.2 SWITCHES AND TRANSDUCERS

Some simple sensors can distinguish between only two different states of the measured variable. Such sensors are called switches. Other sensors, called transducers, provide output signals (usually electrical) that vary in strength with the condition being sensed. Figure 4.5 shows the difference in outputs of a switch and a transducer to the same sensed condition.

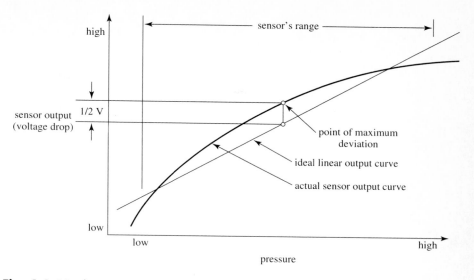

Fig. 4.4 Non-linearity in a pressure sensor

4.2.1 Switches

The most commonly-used sensor in industry is still the simple, inexpensive **limit switch**, shown in figure 4.6. These switches are intended to be used as presence sensors. When an object pushes against them, lever action forces internal connections to be changed.

Most switches can be wired as either **normally open (NO)** or **normally closed (NC)**. If a force is required to hold them at the other state, then they are **momentary contact** switches. Switches that hold their most recent state after the force is removed are called **detent** switches.

Most switches are **single throw (ST)** switches, with only two positions. Switches that have a center position, but can be forced in either direction, to either of two sets of contacts, are called **double throw (DT)**. Most double throw switches do not close any circuit when in the center (normal) position, so the letters "co," for center off, may appear on the spec sheet.

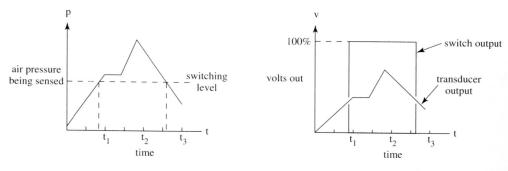

Fig. 4.5 Switch output versus transducer output (switch without hysteresis)

Switches that change more than one set of contacts ("poles") with a single "throw" are also available. These switches are called **double pole (DP), triple pole (TP),** etc., instead of the more common **single pole (SP).**

The limit switches discussed so far all require physical contact between conductive materials inside the switch housing. Electrical arcing across the air gap when the switch is almost closed, and the presence of nonconductive pollutants on the contact surfaces (dirt, oil, corrosion, etc.) may make the dependability of these devices poor. Using the correct-sized switch improves dependability.

In switches designed for high current applications, the contacts are made crude but robust, so that arcing does not destroy them. For small current applications, typical in computer controlled applications, it is advisable to use sealed switches with plated contacts to prevent even a slight corrosion layer or oil film that may radically affect current flow. Small limit switches are often called **microswitches.**

Any switch with sprung contacts will allow the contacts to *bounce* when they change position. A controller monitoring the switch would detect the switch opening and closing rapidly for a short time. Arcing of current across contacts that are not yet quite touching looks like contact bouncing, too. Where the control system is sensitive to this condition, two solutions are possible without abandoning limit switches.

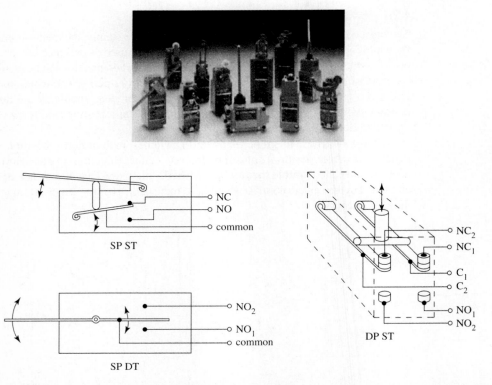

Fig. 4.6 Limit switches. (Photograph by permission, Allen-Bradley Canada Ltd., A Rockwell International Company)

80 4 SENSORS

One solution, used in computer keyboards to prevent single keystrokes from being mistaken as multiple keystrokes, is to include a keystroke-recognition program that refuses to recognize two sequential "on" states from a single key unless there is a significant time delay between them. This method of **debouncing** a switch can be written into machine-language control programs.

Another solution to the bounce problem, and to the contact conductivity problem, is to select one of the growing number of **non-contact limit switches.** These limit switches are not actually limit switches but are supplied in the same casings as traditional limit switches and can be used interchangeably with old style limit switches. Objects must still press against the lever to change the state of these switches, but what happens inside is different.

The most common non-contact limit switch, shown in figure 4.7, is the **Hall effect switch.** Inside this switch, the lever moves a magnet toward a Hall effect sensor. An electric current continuously passes lengthwise through the Hall effect sensor. As the magnet approaches the sensor, this current is forced toward one side of the sensor. Contacts at the sides of the Hall effect sensor detect that the current is now concentrated at one side; there is now a voltage across the contacts. This voltage opens or closes a semiconductor switch. Although the switch operation appears complex, the integrated circuit is inexpensive.

Non-contact limit switches act like switches, even though they include transducers. (The voltage across the Hall effect sensor is proportional to the closeness of the magnet.) We will see in the following sections that there are several switching sensors that are in fact transducers in disguise.

4.2.2 Non-Contact Presence Sensors (Proximity Sensors)

The limit switches discussed in the previous section are "contact" presence sensors, in that they have to be touched by an object for that object's presence to be sensed. Contact sensors are often avoided in automated systems because wherever parts touch there is wear and a potential for eventual failure of the sensor. Automated systems are increasingly being designed with non-contact sensors. The three most common types of non-contact sensors in use today are the **inductive proximity** sensor, the **capacitive proximity** sensor, and the **optical proximity** sensor. All of these sensors are actually transducers, but they include control circuitry that allows them to be used as switches. The circuitry changes an internal switch when the transducer output reaches a certain value.

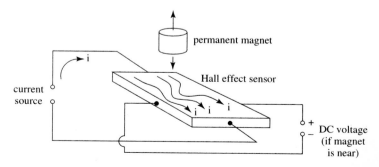

Fig. 4.7 Hall effect limit switches

PNP Versus NPN Sensors

The examples in this section are all PNP type. They show *three contact wires* on transducers used as switches. DC power is supplied to the sensor through the P and N (positive and negative) contacts. The other P contact is the (positive) DC output that is opened or closed when the sensor detects a "target" object. NPN sensors, which have negative DC output, are also available.

Switched proximity sensors are often supplied with *more than three contact wires*, so the same sensor can be connected as PNP or NPN and as either a normally open or a normally closed switch. Some proximity sensors are now available with only *two contacts*. They can be connected like limit switches. The controller supplies DC, just as it would to a limit switch, and reads the switch's output to detect whether it is open or closed. Unlike

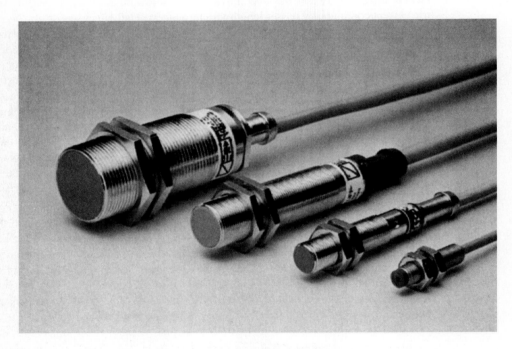

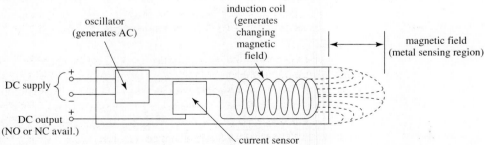

Fig. 4.8 Inductive proximity sensors. (Photograph by permission, Balluff Inc., Florence, Kentucky.)

a true switch, these proximity sensors always consume a small amount of current to operate the internal transducer circuitry.

The **inductive proximity sensor** is the most widely used non-contact sensor due to its small size, robustness, and low cost. This type of sensor can detect only the presence of electrically conductive materials. Figure 4.8 demonstrates its operating principle.

The supply DC is used to generate AC in an internal coil, which in turn causes an alternating magnetic field. If no conductive materials are near the face of the sensor, the only impedance to the internal AC is due to the inductance of the coil. If, however, a conductive material enters the changing magnetic field, eddy currents are generated in that conductive material, and there is a resultant increase in the impedance to the AC in the proximity sensor. A current sensor, also built into the proximity sensor, detects when there is a drop in the internal AC current due to increased impedance. The current sensor controls a switch providing the output.

Inductive proximity sensors, like all proximity sensors discussed here, have a built-in hysteresis. This means that if the sensor's output goes on when an approaching object moves to within (for example) ten millimeters, then it will not go off again until the object moves away to further than (for example) 11 millimeters.

Capacitive proximity sensors sense "target" objects due to the target's ability to be electrically charged. Since even non-conductors can hold charges, this means that just about any object can be detected with this type of sensor. Figure 4.9 (next page) demonstrates the principle of capacitive proximity sensing.

Inside the sensor is a circuit that uses the supplied DC power to generate AC, to measure the current in the internal AC circuit, and to switch the output circuit when the amount of AC current changes. Unlike the inductive sensor, however, the AC does not drive a coil, but instead tries to charge a capacitor. Remember that capacitors can hold a charge because, when one plate is charged positively, negative charges are attracted into the other plate, thus allowing even more positive charges to be introduced into the first plate. Unless both plates are present and close to each other, it is very difficult to cause either plate to take on very much charge. *Only one of the required two capacitor plates is actually built into the capacitive sensor!* The AC can move current into and out of this plate only if there is another plate nearby that can hold the opposite charge. The target being sensed acts as the other plate. If this object is near enough to the face of the capacitive sensor to be affected by the charge in the sensor's internal capacitor plate, it will respond by becoming oppositely charged near the sensor, and the sensor will then be able to move significant current into and out of its internal plate.

Most capacitive proximity sensors can be adjusted to switch at different levels of internal current flow, so that the sensor can be adjusted to sense a wide range of materials, at an adjustable distance.

Capacitive proximity sensors are generally larger and more expensive than inductive proximity sensors. They can give **false readings** if not used carefully. When used to detect the presence of a low density material, it may be fooled by the presence of a distant, higher density material even if the low density part is not present! In one system that the author designed, the sensor reported the presence of a plastic box every time a forklift drove by in the neighboring aisle.

Optical proximity sensors generally cost more than inductive proximity sensors, and about the same as capacitive sensors. They are widely used in automated systems

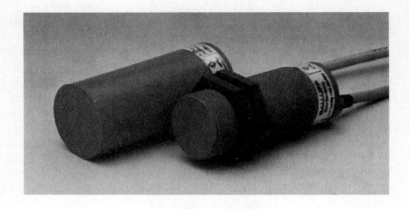

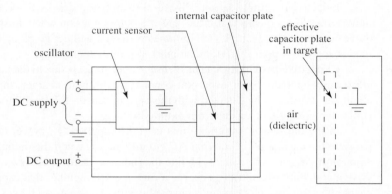

Fig. 4.9 Capacitive proximity sensors. (Photograph by permission, Balluff Inc., Florence, Kentucky.)

because they have been available longer and because some can fit into small locations. These sensors are more commonly known as **light beam sensors** of the **thru-beam** type or of the **retroreflective** type. Both sensor types are shown in figure 4.10.

A complete optical proximity sensor includes a light source, and a sensor that detects the light.

The **light source** is supplied because it is usually critical that the light be "tailored" for the light sensor system. The light source generates light of a frequency that the light sensor is best able to detect, and that is not likely to be generated by other nearby sources. **Infra-red light** is used in most optical sensors. To make the light sensing system more foolproof, most optical proximity sensor light sources **pulse** the infra-red light on and off at a fixed frequency. The light sensor circuit is designed so that light that is not pulsing at this frequency is rejected.

The **light sensor** in the optical proximity sensor is typically a semiconductor device such as a photodiode, which generates a small current when light energy strikes it, or more commonly a phototransistor or a photodarlington that allows current to flow if light strikes it. Early light sensors used photoconductive materials that became better conductors, and thus allowed current to pass, when light energy struck them. All these devices suffer from a delay between when light is received and current is allowed to pass. The semiconductor devices have the least delay. As a rule, the more current the sensor controls, the slower the sensor responds.

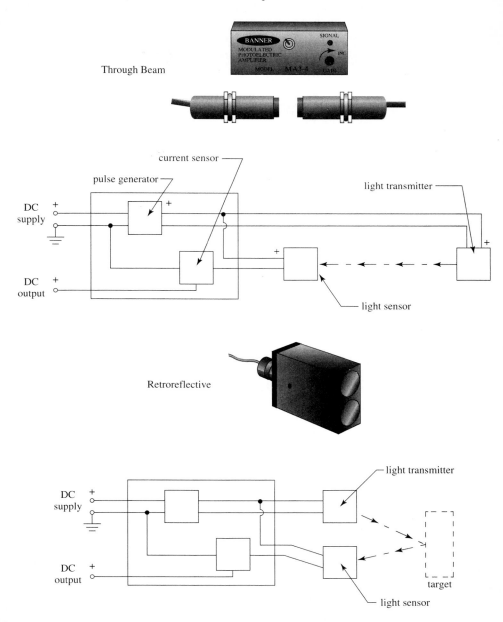

Fig. 4.10 Optical proximity sensors. (Photographs by permission, Banner Engineering Corporation, Minneapolis, Minnesota.)

Sensor control circuitry is also required. The control circuitry may have to match the pulsing frequency of the transmitter with the light sensor. Control circuitry is also often used to switch the output circuit at a certain light level. Light beam sensors that output voltage or current proportional to the received light level are also available.

Through beam type sensors are usually used to signal the presence of an object that blocks light. If they have adjustable switching levels, they can be used, for example, to detect whether or not bottles are filled by the amount of light that passes through the bottle.

Retroflective type light sensors have the transmitter and receiver in the same package. They detect targets that reflect light back to the sensor. Retroreflective sensors that are focused to recognize targets within only a limited distance range are also available.

4.2.3 Temperature Transducers

In the previous section, we discussed several transducer-type sensors that were used as switches. Temperature transducers are almost always used as transducers. Figure 4.11 shows the four types of temperature sensors we will examine.

Probably the most common temperature sensor is the metal **RTD**, or **Resistive Temperature Detector**, which responds to heat by increasing its resistance to electric current. The **thermistor** type of temperature sensor is similar, except that its resistance decreases as it is heated. In either case, there is only a tiny variation in current flow due to temperature change. Current through an RTD or thermistor must be compared to current through another circuit containing identical devices at a reference temperature to detect the change. The freezing temperature of water is used as the reference temperature.

Semiconductor **integrated circuit** temperature detectors respond to temperature increases by increasing reverse-bias current across P-N junctions, generating a small but detectable current or voltage proportional to temperature. The integrated circuit may contain its own amplifier.

Thermocouple type temperature sensors generate a small voltage proportional to the temperature at the location where dissimilar metals are joined. The reason a voltage is generated is still a source of debate. One possible reason may be that heat causes electrons in metals to migrate away from the heated portion of the conductor, and that this tendency is greater in one of the metals than in the other.

4.2.4 Force and Pressure Transducers

Pressure is a measure of force as exerted by some elastic medium. Compressed air exerts a force as it tries to return to its original volume. Hydraulic fluids like oil, although considered "incompressible" in comparison to gases, do in fact reduce in volume when pressurized, and exert force as they attempt to return to their original volume. **Pressure sensors** measure this force.

All pressure sensors measure the difference in pressure between two regions by allowing the pressure from one region to exert its **force** against a surface sealing it from a second region. In most pressure sensors, this second region is the ambient pressure. Pressure **gauges** measure how far the pressure from the measured region moves a load that is held back by room air pressure. Some pressure sensors have two controlled pressure inlets so that the secondary pressure inlet can be connected from a second pressure source. These pressure sensors are called **differential** pressure sensors.

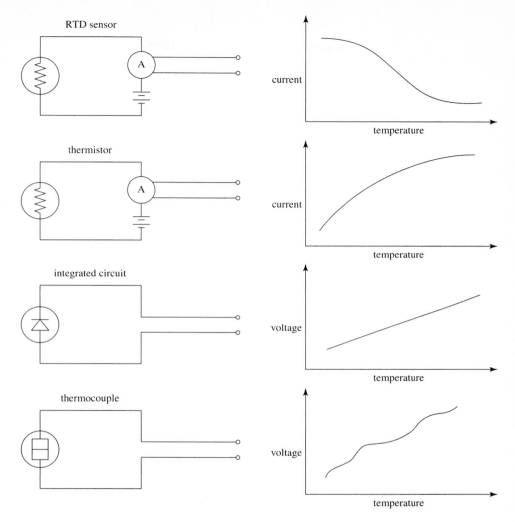

Fig. 4.11 Temperature sensors

Quite a few force sensors are **spring-type** devices, where a spring is compressed by the force. A position sensor detects how far the spring compresses, and therefore how much force caused that compression. Figure 4.12 (next page) shows how this type of force sensor might be built. Position sensors are discussed later in this chapter.

Strain gauges are sensors that measure deformation due to pressure. Figure 4.13 shows that a strain gauge is essentially a long thin conductor, often printed onto a plastic backing in such a way that it occupies very little space. When the strain gauge is stretched, the conductor reduces its cross-sectional area and thus can carry less current. The change in resistance is small and so requires a reference resistance and compensating circuitry to compensate for other sources of resistance changes (such as temperature!). Strain gauges

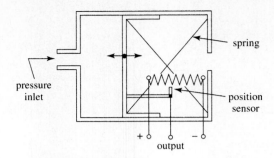

Fig. 4.12 A spring-type pressure sensor

are often glued to critical machine components to measure the deformation of those components under loaded conditions.

The **piezoelectric** strain sensor includes a crystalline material that develops a voltage across the crystal when the crystal is deformed. The small voltage requires amplification. Piezoelectric crystals are often used in accelerometers to measure vibration. Accelerometers will be discussed in a later section.

Bubbler type sensors are used to measure pressure (and therefore depth) in liquids. Air is forced into a tube at a fixed flow rate. The other end of the tube is in the liquid, so bubbles form as the air exits the tube. If the liquid at the bubble end of the tube is under high pressure, more air pressure is required to maintain the constant airflow rate. A pressure sensor in the air line effectively measures the liquid pressure.

4.2.5 Flow Transducers

Flow sensors measure the volume of material that passes the sensor in a given time. Such sensors are widely used in process control industries. Figure 4.14, p. 90, demonstrates the principles used in several types of flow sensors.

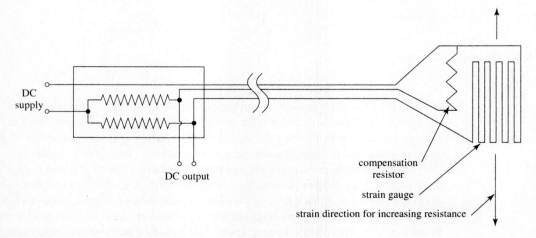

Fig. 4.13 Strain gauges and compensation circuitry

Pressure sensors are often used to measure fluid flow. The faster the fluid is flowing, the more pressure it will create in the open end of the **pitot type** flow meter. Pressure upstream from a **restricted orifice** in a pipe is always higher than pressure downstream from that restriction. The greater the flow rate, the greater the pressure difference, so if a **differential pressure** sensor compares pressures before and after the restriction, then flow rate can be determined. The restriction orifice required by such a sensor reduces the flow.

Temperature sensors are used to measure the flow rate of cool liquids. Quickly flowing fluids can cool a sensor in the stream more than slowly moving fluids can. A heat source keeps the sensor within its operational range. Temperature sensors can be small, so flow restriction is minimal.

Vortex shedding flow sensors offer only a small obstruction to the flow. A vortex shedder obstruction creates swirling areas of low pressure behind and on either side of the obstruction as fluid flows past. The faster the fluid flow, the faster these vortexes are created and destroyed. A vibration sensor on the obstruction measures the frequency at which the obstruction vibrates as the vortex location alternates from one side to the other.

Some flow meters operate by measuring the speed at which the flowing fluid causes a shaft to rotate. (Speed sensors are discussed later.) **Vane** type flow sensors are quite popular, because they provide less resistance to flow than many restricted orifice sensors. Flow sensors that allow the flow to rotate a **turbine** provide even less resistance to the flow being measured. The blades of turbine flowmeters are often magnetized, so that a magnetic sensor outside the pipe can detect the rotation speed of the turbine. Although the output rotation speed of both these flowmeters is theoretically proportional to the flow rate, several factors (leakage, viscosity of the fluid, etc.) can affect the precision.

The **nutating disc** flowmeter is very precise. As fluid flows into one chamber of the flowmeter, it builds up pressure, eventually causing the disc to flop. The other chamber then fills as the first chamber releases its contents to the flowmeter outlet. The disc turns a shaft as it flops back and forth. The shaft may also be connected to a mechanical counter, or to an electrical pulse generator, for precise volume measurement.

Magnetic flowmeters can detect the rate of flow of conductive materials. A magnetic field is generated by the sensor, and the conductive material flows through that field. When a conductor moves through a magnetic field, a voltage is induced in the conductor. The greater the speed, the greater the induced voltage. Electrodes on the sides of the pipe measure the voltage induced in the conductive fluid.

4.3 POSITION SENSORS

Many of the presence detectors already discussed are position sensors in that they measure the presence or non-presence of an object at a certain position. This section is devoted to sensors designed to measure *where,* within a range of possible positions, the target actually is.

Arrays of many switches can be used to find an object's actual position. Sensor array position sensors can be expensive because the cost of the sensor must include the cost of the switches, the cost of the connecting circuitry, and the cost of the controller required to evaluate the many signals from the switches. Arrays of switches can be economical if they are mass produced.

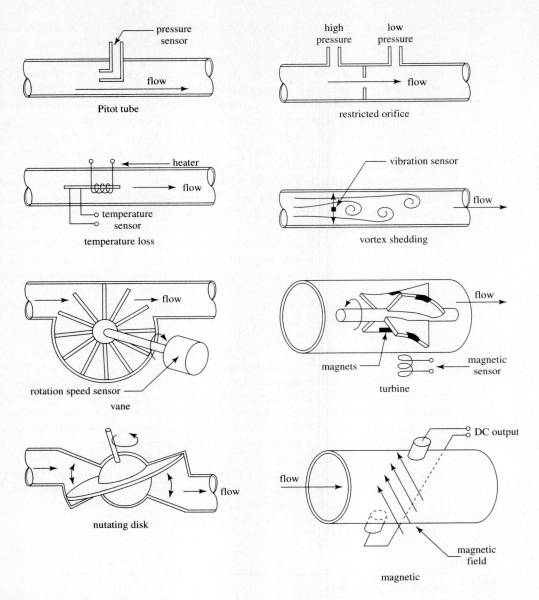

Fig. 4.14 Flow rate sensors

A single housing containing several contacts was once the preferred type of rotary shaft position sensor. The rotary position of the shaft could be determined by counting the number of conductive tracks touching the rotating wiper (see figure 4.15 (a)). The absolute encoder, a variation on this type of position sensor, is very popular and will be discussed later in this section.

Switches built into arrays in a single integrated circuit or printed onto a circuit board are other types of switch arrays that are economically viable.

Photodiodes, printed only millimeters apart in linear arrays on metal surfaces as shown in figure 4.15 (b), are in use as position sensors. Columnated light is projected onto the photosensor array, and the location of the target can be determined by examining which of the photodiodes is not conducting because it isn't receiving light.

Pressure switch arrays (figure 4.15 (c)) are available. They enable automated systems to find the location and orientation of parts resting on the switch array. (Pressure transducer arrays use the amount of pressure at each sensor in their evaluation programs.)

CCD cameras (see chapter 5) include light sensor arrays on a single IC chip. Each sensor is usually used as a transducer, not as a switch.

As the position-sensing task using switch arrays becomes more complex, the need for computerized analysis becomes greater, so most suppliers of switch arrays also supply controllers.

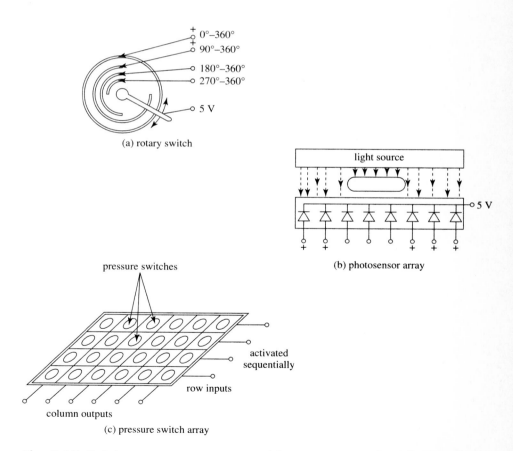

Fig. 4.15 Switch arrays as position sensors: (a) rotary position sensor; (b) photodiode array; (c) pressure switch array.

4.3 Position Sensors

The majority of position sensors are not switch arrays, but **transducers,** or sometimes transducer arrays.

The **inductive** and **optical** transducers described in the section on non-contact presence sensors are available for use as position sensors. Instead of control circuitry containing switched output, when used as position sensors they must include control circuitry to output an analog value (e.g., voltage or current) linearly proportional to the sensed distance from the transducer to the "target." Since the inductive and optical transducer's output is exponentially related to target distance, the sensor control circuitry must translate the exponential relationship to a linear one. A typical inductive range sensor would have a range of about 15 mm, while an optical position sensor could have a range as long as a few inches.

Potentiometers, also called "pots" or variable resistors, are making a comeback as position sensors. Pots, shown in figure 4.16, can be used as linear or as rotary position sensors. Pots output a voltage proportional to the position of a wiper along a variable resistor. They have a reputation for being too inaccurate because of the effects of temperature changes, wear, and resistance changes due to pollutants between the wiper and the variable resistor. Pots are making a comeback because developments in materials, especially in conductive plastics, are improving their performance.

A dependable position sensor with an intimidating name is the **linear variable differential transformer,** or **LVDT.** Shown in figure 4.17, the LVDT is provided with AC at its central (input) coil. The transformer induces AC into the output coils above and below this input coil. Note that the two output coils are connected in series and are opposite wound. If the core is exactly centered, the AC induced into one output coil exactly cancels the AC induced in the other, so the LVDT output will be 0 VAC. The transformer core is movable in the LVDT housing. If the core lifts slightly, there is less voltage induced in the lower output coil than in the upper coil, so a small AC output is observed, and this output voltage increases with increased upward displacement of the core. If the upper coil is wound in the same direction as the input coil, this output voltage is in phase

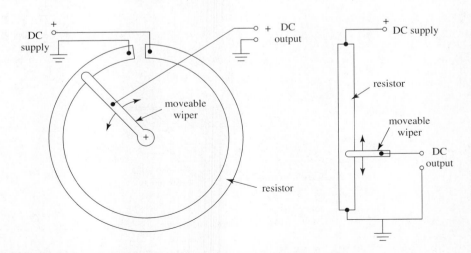

Fig. 4.16 Potentiometric position sensors

with the input AC. If the core displaces downward from center, output voltage will also increase proportionally with displacement, but the output AC waveform will be 180 degrees out of phase.

Magnetostrictive position sensors are recent arrivals as position sensors. They detect the location of a magnetic ring that slides along a conductive metal tube. A magnetostrictive position sensor is shown in figure 4.18 (next page). To detect the position of the magnet, a pulse of DC current is introduced into the tube. Some time later, the current pulse reaches the magnet and passes through its magnetic field. When current moves across a field, the conductor experiences a force. The tube distorts, and a vibration travels back along the tube to a force sensor. The time elapsed between generation of the DC current and the time of the vibration is linearly related to the position of the magnet along the tube.

Capacitive position sensors have been used as dial position sensors in radios for years. (Their variable capacitance is used in the radio frequency selection circuitry, so perhaps calling them "sensors" in this application isn't quite right.) A capacitor increases in capacitance as the surface area of the plates facing each other increases. If one 180 degree set of plates is attached to a rotating shaft, and another 180 degree set of plates is held stationary as demonstrated in figure 4.19 (next page), then capacitance increases linearly with shaft rotation through 180 degrees.

Capacitive sensors are used in a slightly different way to detect the height of fluids that have higher dielectric constraints than air. If the fluid rises in the capacitive height sensor shown in figure 4.19, then more surface area of the capacitor plates are coupled by the fluid, and capacitance rises proportionally with fluid height.

A wide assortment of position sensors work on the principle of **reflected waveforms.** Several reflected waveform principles are shown in figure 4.20 (p. 96). The simplest of this category is the **retroreflective light beam sensor** (previously discussed, but

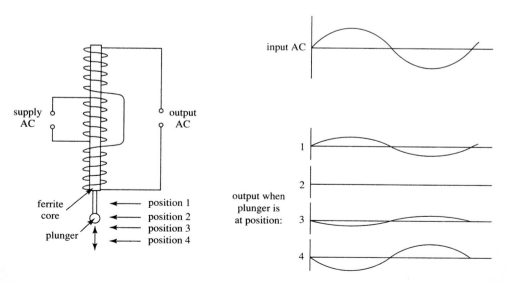

Fig. 4.17 The linear variable differential transformer (LVDT)

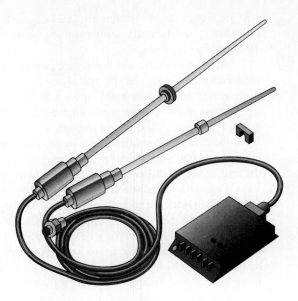

Fig. 4.18 The magnetostrictive position sensor. (By permission, Deem Controls Inc., London, Ontario, Canada.)

this time using the transducer's analog output). The sensor's output is proportional to the amount of light reflected back into the light detector, and therefore proportional to the nearness of the reflective surface.

Just slightly more complex are the sensors, such as **ultrasound scanners,** in which a short pulse of energy (in this case, high frequency sound) is generated. The distance to the target is proportional to the time it takes for the energy to travel to the target and to be reflected back. As used in medical ultrasound scanners, a portion of the energy is reflected

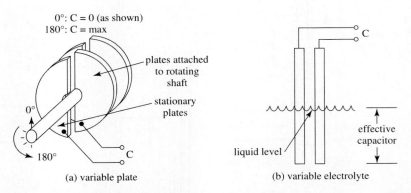

Fig. 4.19 Capacitive position sensors: (a) variable plate engagement; (b) variable electrolyte presence

by each change in density of the transmission media, and multiple "target" locations can be found. Ultrasound is now available in inexpensive sensors to detect distances to solid objects.

More sophisticated and much more precise location measurements can be done with **interferometer type sensors,** which use energy in the form of (typically) light or sound. The transmitted wave interacts with the reflected wave. If the peaks of the two

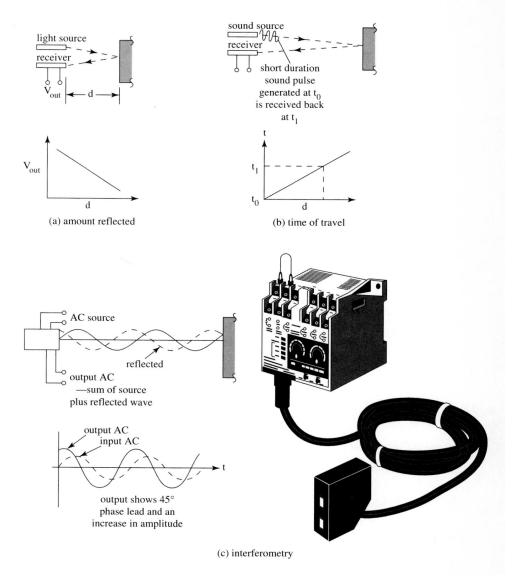

Fig. 4.20 Reflected waveform sensors: (a) amount of reflected waveform; (b) time of travel; (c) interferometry. (Photograph of ultrasonic sensor by permission, OMRON Canada Inc., Scarborough, Ontario, Canada.)

waveforms coincide, the resultant waveform amplitude is twice the original. If the reflected wave is 180 degrees out of phase with the transmitted wave, the resultant combined waveform has zero amplitude. Between these extremes, the combined waveforms result in a waveform that is still sinusoidal, but has an amplitude somewhere between zero and twice that outputted, and will be phase shifted by between 0 and 180 degrees. This type of sensor can determine distance to a reflective surface to within a fraction of a wavelength. Since some light has wavelengths in the region of 0.0005 mm, this leads to a very fine precision indeed. If **laser** light is used, the waveforms can travel longer distances without being reduced in energy by scattering.

Some position sensors are designed to measure the *rotational position* of shafts. Two similar types of shaft rotational position sensors are the **rotary resolver** and the **rotary synchro.**

The **resolver** is slightly less complex, so we will discuss it first. Figure 4.21 shows that in construction the resolver looks very similar to a DC motor. It has field windings (two of them, at 90 degrees to each other) and a winding in the rotor. The rotor winding is electrically connected to the outside world through slip rings (not a commutator). In operation, the resolver has more in common with a transformer than it does with a motor. The following description will examine only one possible connection method.

An AC supply is connected to the winding in the rotor. The rotor is rigidly connected to the rotating shaft. When the rotor is in the zero degree position, it induces AC in the cosine field winding, but not in the sine winding. As the rotor moves from the zero degree position toward the 90 degree position, it induces progressively less AC in the cosine winding, and progressively more AC in the sine winding, until at 90 degrees, there is no AC induced in the cosine winding, and the AC output from the sine winding is at a maximum. As the rotor moves from 90 degrees to 180 degrees, the AC output from the sine winding reduces to zero again, and the AC output from the cosine winding increases back to a maximum, but this time at 180 degrees out of phase with the supply AC.

Synchros are different from resolvers in that they include a third "field" winding. The three "field" windings are at 120 degrees from each other. This extra winding allows synchros to be used where added precision is required.

Suppliers of resolvers and synchros also supply **interfacing circuits** that analyze the output waveforms for the user. The analyzer output is usually a DC value proportional to the shaft angle. The resolution of these position sensors is limited only by the quality of the sensor and analyzer components. The cost of a resolver or a synchro is often higher than can be justified.

Optical encoders are perhaps the most common shaft position sensor used today. As we will see, they are ideally suited for digital controller use. They come as either absolute or as incremental encoders. Of the two, the incremental encoder is most widely used, so we will discuss it first.

The **incremental optical encoder,** shown in figure 4.22, consists of a light source, one or two disks with opaque and transparent sections, three light sensors, and a controller. In the single disk system (shown in the diagram), the disk is attached to the rotating shaft. The stationary light sensors detect light when the transparent section of the disk comes around. The encoder's controller keeps track of the shaft position by *counting the number of times the sensors detect changes in light.*

Resolution of the sensor increases with the number of transparent sections on the disks, so a two-disk system is more common than the single-disk system. In this type, both disks have finely etched radial transparent lines, as demonstrated with the upper incremental disk in Figure 4.22 (next page). One disk is held stationary while another is attached to the rotating shaft. The sensor will thus see light only when the transparent sections on both disks line up.

The controller must also detect the *direction of rotation* of the shaft. It keeps track of the rotary position by adding or subtracting from the last position every time light is received. Figure 4.22 shows a second track of transparent sections, at 90 degrees from the first. If the shaft is rotating in a clockwise direction, the outer light sensor sees light first. If rotating in a counterclockwise direction, the inner light sensor sees light first.

The third, inner sensor is used to *initialize the count*. When the light source is unpowered, motion of the shaft goes undetected, so position control systems using incremental encoders must always be "homed" after power has been off. The user first rotates the encoder shaft until it is within one revolution of its "home" position. The controller is then sent a signal that tells it to watch for output from the inner sensor and initialize its counter value when the sensor detects light. The shaft is then slowly rotated. When the one transparent section of the disk at the third sensor comes around, the count is set to zero.

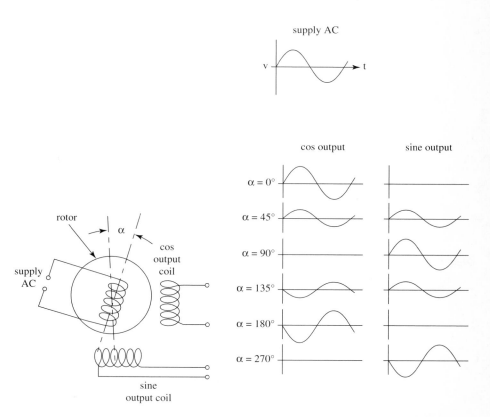

Fig. 4.21 The rotary resolver

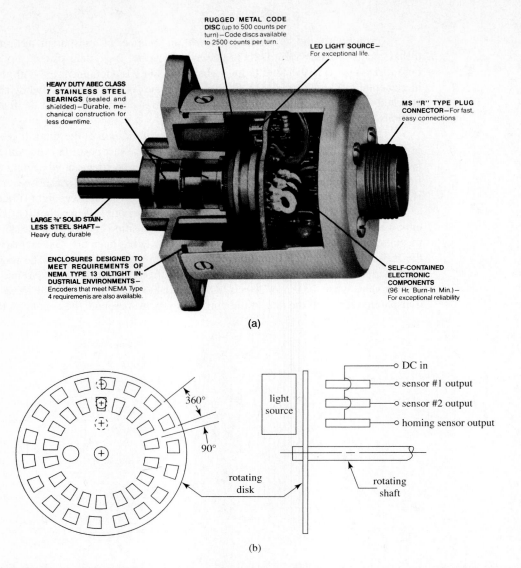

Fig. 4.22 The incremental optical encoder. (a) Cutaway view of encoder; (b) the etched disk. (Cutaway view of encoder by permission, Allen-Bradley Canada Ltd., A Rockwell International Company. Etched disk by permission, Parker Hannifin Corporation, Compumotor Division, Rohnert Park, California.)

The initiation circuit is then disabled and the counter proceeds to increment and decrement as the other two light sensors detect shaft rotation.

The incremental encoder must at least include circuitry to cause the light transducers to act as switches. Nothing more is needed if a digital controller is programmed to initialize and keep track of the count. Optionally, the encoder supplier may include sufficient built-in features to initialize and maintain the count, so that a digital controller need only read the position when required. Some controllers are capable of providing an output signal at a preset position.

Unlike incremental optical encoders, **absolute optical encoders** do not require homing. These encoders consist of a light source, a rotating disk with more than three circumferential sets of transparent sections, a light sensor for each ring of slots, and a circuit card.

Absolute optical encoders are used to detect position of a rotating shaft within a *single* revolution. Absolute encoders output a binary number representing the position of the disk (which is rotated by the shaft). This binary number has as many bits as there are light sensors and rings in the disk. If the light sensor receives light, it outputs a "1"; otherwise it outputs "0." Absolute optical encoder disks are shown in figure 4.23. The diagrams show hypothetical disks with four rings of transparent sections and a more typical nine-ring disk.

If a natural **binary** numbering system is used, the inner light sensor provides the most significant binary bit (as shown in figure 4.23). The transparent section at the inner ring on the disk extends for 180 degrees. The second most significant bit is received from the second sensor from the center, which sees light through a ring of two 90 degree long slots. These first two sensors output the binary number 00 if the disk is rotated at between 0 and 90 degrees, 01 if the disk is rotated to between 90 and 180 degrees, 10 if between

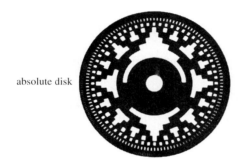

absolute disk

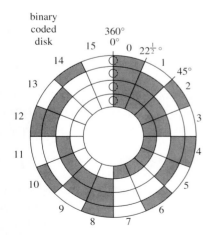

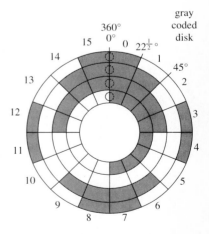

zone	binary	gray
0	0000	0000
1	0001	0001
2	0010	0011
3	0011	0010
4	0100	0110
5	0101	0111
6	0110	0101
7	0111	0100
8	1000	1100
9	1001	1101
10	1010	1111
11	1011	1110
12	1100	1010
13	1101	1011
14	1110	1001
15	1111	1000

Fig. 4.23 Absolute optical encoder. (Photograph by permission, Parker Hannifin Corporation, Compumotor Division, Rohnert Park, California.)

180 and 270, and 11 if between 270 and 360 degrees. Each additional ring adds another bit to the binary number, and thus doubles the resultant resolution.

Accuracy is limited by how precisely the slots are positioned, and how precisely the light sensors switch when the slots begin and end. Use of a natural binary numbering system makes erroneous position signals possible. When a four-sensor absolute encoder disk is rotated from segment 15 (binary:1111) to segment 0 (binary:0000), any one of the light sensors could change before the others. In fact, if the computer controller "read" the sensor's output while the disk was exactly at this changeover position, it might get *any* four bit number!

The **Gray** numbering system is used to prevent this type of large error in encoder output. In the Gray numbering system, only one sensor changes its value between one range and the next, so the potential error due to imprecision in transparent section placement or sensor switching times is minimized. Optional circuit boards are available to translate Gray binary numbers to natural binary numbers.

Even if power has been lost, the absolute optical encoder always shows the actual shaft position once power is restored, so it is not necessary to initialize them. High resolution absolute encoders are expensive, however, because of the precision required in manufacturing, and because of the disk size and extra sensors required to increase resolution. Absolute encoder disks should not be allowed to rotate more than a single revolution, or they will require a homing procedure at power up.

Although *optical* encoders have been discussed because they are so common, incremental and absolute encoders can be and are manufactured using *non-optical* switches.

4.4 VELOCITY AND ACCELERATION SENSORS

Velocity is the first differentiation of position, and acceleration is the second. **Position sensors** can be used, therefore, in speed and position control systems.

If the controller is a digital computer, *the change in position per time interval* (first differential of position) can be calculated and used in a speed control program. A change in velocity per time interval calculation (second differential) can be used in an acceleration control program.

The *differentiation can be done in hardware* in an analog controller. Figure 4.24 demonstrates how op amp differentiators can be used to generate velocity and acceleration signals when a simple potentiometer rotary position sensor is attached to a shaft. Op amp differentiators are discussed in chapter 7.

Certain sensors, such as incremental encoders and **magnetic sensors,** emit pulses as they detect positional changes. Magnetic sensors, shown in figure 4.25, consist of magnets embedded in the moving object, and stationary coils. The magnets might be in the vanes of a turbine flowmeter or in the teeth of a gear. As the magnets move past the coils, they induce a voltage in the coil. A controller can *count these pulses* during a time interval to determine speed.

AC generators can be used as speed sensors in high precision control systems. The AC generator rotor is rotated by the rotating shaft. Output AC frequency is proportional to

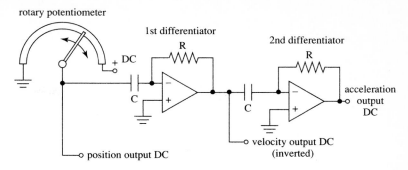

Fig. 4.24 Position sensor and op amps for velocity and acceleration output

shaft speed. Zero crossings per time interval can be counted to determine speed for a simple speed control system.

If high precision speed *and position* control are required, the controller can generate a reference AC with frequency proportional to the desired speed. The control system can monitor the **phase shift** between the reference AC and the sensor's output AC. If the phase shift comparator detects that the actual system is lagging the reference, *even if the speed is the same,* the actuator is driven harder until it catches up with the reference AC. Such a control system is called a **phase locked loop speed control** system. An AC generator and a phase locked loop control system is demonstrated in figure 4.26.

A DC generator sensor (with a commutator) is popularly called a **tachometer.** It generates DC proportional to speed, and can be used as a speed sensor.

Doppler effect speed sensors are used in police radar speed detectors and similar electromagnetic wave speed sensors. Figure 4.27 demonstrates that when a wave of a given frequency reflects from a moving object, the frequency shifts. If the object is moving toward the transmitter/receiver, the frequency increases, and the amount of frequency shift is directly proportional to the approach speed. A receding reflective surface shifts the frequency downward.

Devices to measure acceleration are known as **accelerometers.** Accelerometers measure the force required to cause a mass to accelerate. The housing of the accelerometer shown in figure 4.28 is rigidly attached to the object that is being accelerated. Inside the

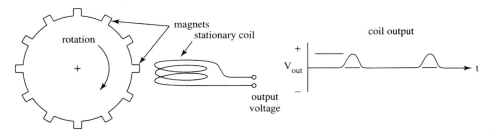

Fig. 4.25 Magnetic sensor

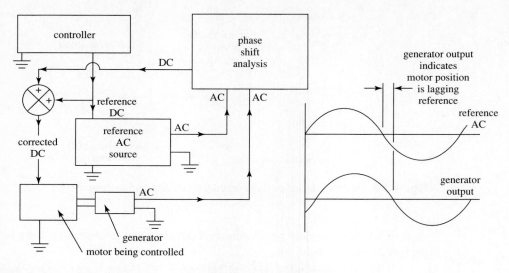

Fig. 4.26 AC generator speed sensor and phase locked loop speed control

housing, a known mass is centered by an arrangement of springs. Because of its inertia, the mass does not accelerate as fast as the housing, so there is a displacement of the mass from the accelerometer center. The amount of displacement, which is proportional to the acceleration moving the housing, can be measured with a position sensor.

Accelerometers may be used as **vibration sensors.** Where high frequency vibrations (such as sound) are measured, spring mass accelerometers may not be fast enough. The springs are replaced by piezoelectric materials that output a voltage proportional to forces that cause only tiny deformations.

4.4.1 Sensors Measuring Alternating Signals

Choosing an accelerometer, or indeed any sensor, to operate in a vibrating system requires that the user be aware of the frequency response of the sensor.

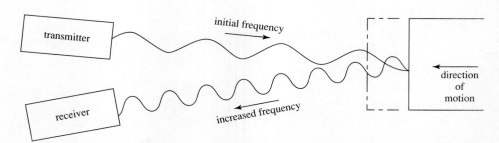

Fig. 4.27 The Doppler effect

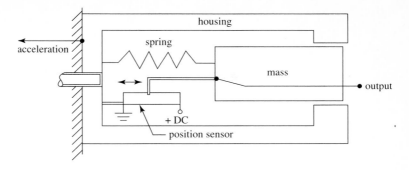

Fig. 4.28 Accelerometers

The Bode diagram for an accelerometer in figure 4.29 shows that, up to a certain frequency (f_1), the output amplitude of the sensor is proportional to the amplitude of the vibration being sensed, and in phase with the vibration. Accelerometer output is error-free.

Near the frequency f_n (the "natural" or **resonant frequency** of the sprung mass in the accelerometer), the sensor's output is larger than it should be.

Above the sensor's resonant frequency, the output amplitude of the sensor will be less than it should be, and will continue to drop off with higher input frequencies until at some frequency (f_2), the sensor will have no output at all. The sensor's output signal will also lag the measured vibration above f_n. (Bode diagrams will be seen again in the section on servosystem response.)

All sensors are subject to **frequency response limitations.** A position sensor that must report the position of a load moving forward and backward at a high frequency, and that is elastically attached to that load, will have a resonant frequency at which and beyond

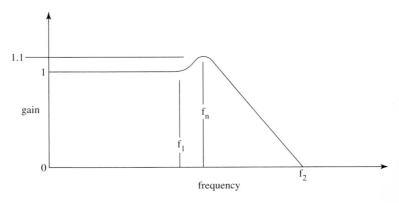

Fig. 4.29 Frequency response of an accelerometer (Bode diagram)

which its output cannot be trusted. Even a temperature sensor requires a short time to respond to a temperature, and if temperature changes too quickly, the sensor will not report the temperature correctly. As a general rule, sensors and sensor systems *should not be required to measure variables that change faster than one half of the natural frequency* (f_n) *of the sensor or system.*

4.5 SPECIAL-PURPOSE SENSOR SYSTEMS

Sometimes a sensor, or a sensor system, is required to output more than just one simple signal, such as for position or velocity. In an earlier section of this chapter, we discussed arrays of switches used to detect object position and orientation. The switch arrays are commonly pre-assembled and sold with a controller capable of monitoring and interpreting the combinations of switch inputs. The controller includes a microprocessor and ROM-based programming for data interpretation, and usually also includes an RS 232C serial port for exchanging data with other computers. The data might be a message from this switch controller, describing the sensed condition, or it might be programming data being sent to the switch controller by the user. The ability of this controller to exchange complex messages makes it more useful in a **Flexible Manufacturing System (FMS)** or in a **Computer Integrated Manufacturing (CIM)** application.

Within an automated manufacturing system, there are often **automated gauging** sensor systems. A data logger can record data from these sensor systems to keep a log of the product quality. If the data is read by an FMS or CIM controller it can even be used in **Statistical Process Control (SPC)** calculations. The SPC results may be used to modify the manufacturing process to ensure high quality.

A **Coordinate Measuring Machine (CMM)** is a computerized system built around position sensors. The CMM's controller monitors several position sensors and a touch sensor. When the touch sensor touches a part, the cartesian (X, Y, and Z) coordinates of the touch sensor's position are read. This data may simply be printed, or might be compared with expected values and/or transmitted to another computer. "Expected" dimensional values may have been received from a Computer Aided Design (CAD) database.

CMMs are most commonly used in the quality control department, where batches of products are inspected after they have been through a manufacturing process, but CMMs can be distributed throughout an FMS to monitor product quality as products are being manufactured. The sensed dimensional data (or deviation from expected data) might be printed out, stored, or used in Statistical Process Control (SPC) calculations.

A **bar code reader,** like that shown in figure 4.30, is another computerized system. The bar code reader is built around retroreflective optical sensors. The light sensors detect differences in light reflected back from bar codes (black marks on lighter areas).

There are two types of **sensor configurations.** The light transmitter and receiver may be combined in a pen-like device called a "wand" (in front of keyboard in figure 4.30). Marks are read by touching the wand to the bar-coded surface and moving the wand past the code. A "scanner" (pistol-grip item in figure 4.30) does not need to touch the bar-coded surface. Stationary scanners have light sources, usually lasers, which are

Fig. 4.30 Bar code reader system, and a typical bar code. (Bar code reader system by permission, Allen-Bradley Canada Ltd., A Rockwell International Company.)

reflected by a moving mirror system to sweep an area in a repeating pattern. Some scanners can sweep a target area hundreds of times a second.

The controller monitors light reflected back to a photodetector, notes the times at which the edges of dark areas begin and end, and evaluates the pattern of observed light/dark regions. If a bar code is detected, the human or computer supervisor is notified, and the code is outputted to a display, or to a recorder, or to another computer.

Bar code specifications require a "quiet zone" of no marks in front of the black marks. The code itself contains coded data between start and stop patterns. Both the width of the dark areas and the width of the light areas are significant. There are several standard codes, perhaps the most-used of which is the commercially-used **Universal Product Code (UPC).** The UPC includes codes for numbers only. Code 39, also known as **Automatic Identification Manufacturers Uniform Symbol Description 2 (AIM USD-2)** is more common in industry. It includes codes for alphanumeric characters and for some control symbols.

Bar codes on materials and tools allow CIM control of the manufacturing process. The shop floor control (SFC) aspect of Manufacturing Resource Planning (MRP II) software requires that the plant's host computer monitor the flow of materials and tools. If each batch of material or work-in-process (WIP) and each manufacturing tool has a bar code affixed, the codes can be read during automated transport, or can be scanned by humans with bar code readers, as they are moved or worked on. The manufacturing-control computers then always have up-to-the-minute information on the state of the manufacturing process. This data can be used to dynamically reschedule manufacturing as soon as the need to do so is recognized.

While bar codes are most commonly used to identify the object carrying the code, they can be used wherever a computer requires data input. Since error rates in reading bar codes are far lower than with some data entry methods, such as keyboard entry, bar codes are sometimes prepared containing machine operator commands. The operator uses a prepared table of command bar codes and a bar code reader to send commands to a machine.

4.6 SUMMARY

We have seen terms describing sensor quality, and can distinguish between **range** and **resolution** and recognize the meanings of terms such as **repeatability, hysteresis, linearity,** and **accuracy** when used to describe sensor characteristics.

We recognized that **switches,** used only as on/off type sensors, are different from **transducers** that output signals that vary in strength as the variables they measure change. In later sections we discussed transducers used as switches, and arrays of switches used in transducer-type applications.

We looked at a wide range of sensors, starting with simple mechanical **limit switches** and **microswitches,** as well as the **Hall effect** switches now sometimes replacing mechanical limit switches.

We looked at how inductive, capacitive, and optical **non-contact proximity switches** work, and discussed the PNP, NPN, and two wire contact types for such switched transducers.

We examined the operating principles of RTD, thermistor, thermocouple, and semiconductor **temperature sensors,** and the operating principles of strain gauge and piezoelectric **pressure and strain sensors.**

Many **flow sensors,** including pitot tube, restricted orifice, and vortex shedding, make use of pressure or strain sensors. One flow sensor uses a heater and a temperature sensor. Many more flow sensors, such as the vane type, the turbine, and the nutating disk flow meters, require shaft position or speed sensors. We examined one sensor, the magnetic flowmeter, which does not impede the fluid flow at all.

Position sensors include a wide range of styles. Arrays of switches can be used, if there is sufficient capability in the controller to read many switches, or if a dedicated controller is used. The vast majority of position sensors are transducers, from the simple resistive potentiometer, to the variable capacitors through to the inductive sensors such as the LVDT, magnetoresistive sensor, resolver, and the synchro. Some position sensors are based on various ways of using reflected waves, including light sensors, ultrasound sensors, and the more sophisticated interferometer sensors.

In **incremental and absolute position encoders** (often optical), light transducer outputs are switched, and a digital computer either keeps track of the number of light pulses (incremental) or reads the pattern of switches (absolute).

Some **velocity sensors** are actually position sensors, and the digital computer or dedicated analog circuit calculates rate of change of position. **Acceleration sensors** can also be simulated by calculating rate of change of velocity. Magnetic sensors are used as speed sensors. The pulses outputted by the sensor are counted over a known interval. True speed sensors do exist. DC generators (also known as tachometers) and AC generators output signals proportional to rotational speed.

Accelerometers measure the force required to accelerate a known load, and **vibration sensors** use piezoelectric sensors that output an alternating voltage as they are deformed. If sensors are used to measure rapidly changing variables, the user should be aware that each sensor has a useful range of frequencies, above which their output is not valid.

Some sophisticated sensor systems come complete with computerized controllers that interpret sensory input. Some switch array sensors are built into systems of this type. **Coordinate Measuring Machines** and **bar code readers** include controllers and communications capabilities that make them valuable in CIM applications.

REVIEW QUESTIONS

1. A sensor's specification sheet says that it can measure air pressures between −10 pounds per square inch (psi) and 120 psi. The sensor has a resolution of 0.25 psi, a repeatability of +/−2 psi, and a linearity of +/−5 psi.

   a) What is the input span of this sensor?

   b) If the actual pressure being measured is 60 psi, what is the maximum that the pressure can rise to without the sensor's output changing?

   c) At a pressure of 60 psi, what is the range of possible sensor output values, given the linearity specification? Is the actual output likely to contain this much error at 60 psi? Why or why not?

   d) Assume that the linearity is perfect as the pressure rises, and perfect as the pressure falls. The reading might still be different as the pressure rises through 60 psi compared to the reading as the pressure falls through 60 psi. What is the maximum reading that might occur at 60 psi? Will this reading be observed as the pressure rises or as it falls?

2. Sketch and identify all of the possible contacts on a DPDT co switch.

3. What type of limit switch would you select for very low power applications, if contact contamination has been a problem?

4. Sketch and indicate the purpose of the contacts of a PNP proximity sensor.

5. Describe what happens as a PNP capacitive proximity sensor (NO) detects an approaching object. Clearly identify all of the internal and external signals and their changes.

6. What two reasons are there for the matching of light beam transmitters and receivers?

7. How does the resistance of a thermistor temperature transducer change as the measured temperature rises?

8. How does a vortex shedder type flow sensor detect increasing fluid flow?

9. Which of the flow sensors in figure 4.14 is better for measuring the total amount of fluid that flows past the sensor over a given time interval?

10. What is the output of an LVDT position sensor at zero displacement? How does this output change as displacement increases? What difference would there be in the sensor's output while it measures positive displacement versus when it measures negative displacement? (Use sketches of input and output signals with your answers.)

11. Sketch a method you might use to use a magnetostrictive position sensor to measure liquid height in a tank.
12. What are each of the three rings of holes on an incremental encoder disk for? Sketch how these rings of slots would have to be arranged in rows in a *linear* incremental encoder.
13. How many rings of slots and light sensors would be required for an absolute optical encoder with a resolution of 1° or finer (capable of differentiating among at least 360 different zones of the encoder disk in one revolution)?
14. A controller outputs a DC voltage proportional to the desired DC motor speed (e.g., 5 V for 500 rpm, 10 V for 1000 rpm). A phase locked loop controller generates an AC frequency perfectly proportional to the DC signal. A phase shift comparator compares the reference AC and the AC generated by the AC generator type sensor.
    a) How will the AC from the AC generator change as the motor accelerates?
    b) If the motor is running at the correct speed, but the shaft position lags where it should be by a small amount, how does the phase shift analysis box detect this positional error? Will it output + or − DC to correct the positional error? (The answer is not in the text; you must figure this one out yourself.)
15. A Bode diagram similar to the one in this chapter shows the response of a sensor to varying input frequencies. If f_1 is 1000 Hz, f_n is 1400 Hz, and f_2 is 10,000 Hz, what is the maximum frequency that the sensor can be used to measure?

CHAPTER 5

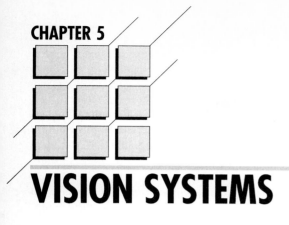

VISION SYSTEMS

CHAPTER OBJECTIVES

After completing this chapter, the reader will be able to:

- Identify, and describe the functions of, vision system components.
- Select techniques for altering images presented to a vision system, making recognition easier.
- Discuss the relative merits of different cameras and lenses for capturing of images. The operating principles of CCD and Vidicon cameras will be understood.
- Discuss the relevance of the following in converting camera images to digital data:
  - sampling
  - quantization
  - grey scale reduction
  - run length coding
  - windowing
- Discuss, in general, several processor schemes used to:
  - enhance, or correct images
  - process data for recognition of two- or three-dimensional object type, location, orientation, or other characteristics
  - analyze scenes to recognize, for example, parts in jumbled bins, or motion of an object
- Discuss templates: How they are created and how they are used in recognition of parts or of part characteristics.
- Describe output requirements for a vision system.

5.0 VISION SYSTEMS

Vision systems are becoming more common in industrial control applications. This is because computer technology has improved to the point that components are now cheap and dependable, and fast enough to keep up with typical manufacturing processes.

Increased robot usage provides a need for vision systems. Robots, with their inherent flexibility, require *sensors that provide lots of information* about the robot's environment, so that the robot's response to that environment can be appropriate. Using standard sensors and dedicating one whole robot input to each sensor would otherwise severely limit the amount of information that a robot could receive.

The modern vision system includes a *dedicated computer that can "process" a scene* and output only the amount of information required by the process control equipment (perhaps a robot). The process control equipment then works out the appropriate response, while the vision system goes on to analyze the next screen of vision information.

The developments in vision system technology are applied with non-optical sensing technologies as well. Much of the data capturing and data processing that will be discussed in this chapter is already finding use in applications such as tactile sensing.

5.1 SIMPLE VISION SYSTEMS

Figure 5.1 shows a system that might be considered an extremely simple vision system. In this example, the system consists of a light source and two light sensors. If there is a tall object in front of the light source, the light reflects into both sensors. If there is a short object in front of the light, the light reflects into only the lower sensor. The "picture" of the object consists of two picture elements: one from each light sensor. "Picture elements" are commonly called **pixels.**

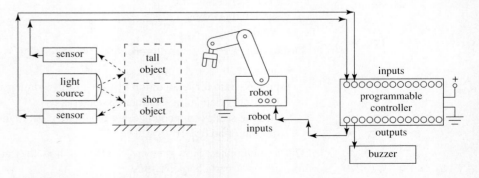

Fig. 5.1 A two pixel vision system

Our two pixel vision system has a computerized part recognition system that analyzes the data received from the sensors and provides outputs depending on the result of the analysis. The system can be an inexpensive programmable controller (see chapter 7). It is programmed to notify a robot if a tall object is observed, or actuate a buzzer if a short object is recognized. There is no output if there is no object in the field of view. This simple vision system saves the robot from having to respond unless a tall object is received, and saves the human operator from being distracted from his or her other duties unless a short object is received.

Calling the system in figure 5.1 a vision system stretches the definition of a vision system slightly. The system in figure 5.2 is definitely a vision system, although still a simple one. This vision system can distinguish between several different objects in its field of view, and can tell the robot *which* of the many possible parts is present. The robot can then react to the presence of that part without having to waste valuable time recognizing the part itself. If the system designer wishes, the vision system can be programmed to do additional image analysis such as quality control analysis of the part, or evaluation of the part's position so the robot can pick it up, etc. For this example, we will assume that simply recognizing the type of part is sufficient.

The simple vision system in figure 5.2 consists of:

- the **light,** reflected by objects in the field of view. Light is received by …
- the **camera,** which converts the light levels to electric charges, which provide current to …
- the **"framegrabber"** circuitry and programming, which breaks the field of view into pixels, and assigns a binary number for each pixel, representing the amount of light received at that pixel. The binary numbers are sent to …

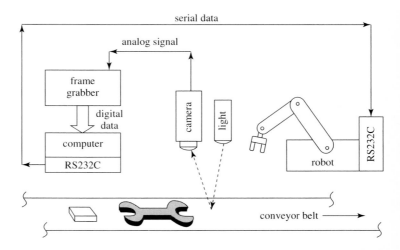

Fig. 5.2 A simple part recognition system

5.1 Simple Vision Systems **111**

- the **computer,** which has memory where the numbers are stored. In this example, the computer counts the number of bright pixels received in a single "frame" from the camera. The sum is proportional to the size of the part on the conveyor belt. Each different part has a different surface area, so the computer can determine which of the parts is in the field of view by determining its area. The computer tells the robot which part has arrived through …

- the **output interface,** which carries message signals from the vision system's computer to the robot's computer. The vision system computer sends the ASCII code for the letters "W" (for "wrench"), or "B" (for "box") via a standard RS 232 interface.

The biggest difference between the vision systems shown in figures 5.1 and 5.2 is in the number of pixels. As the number of pixels increases, having a single sensor that can view many pixels and a fast computer with lots of memory to analyze the increased amount of data becomes necessary.

It is a good rule that a vision system, including the program that recognizes parts, should be *as simple as possible to minimize service requirements and the potential of recognition mistakes.* Time is often a factor, so the less time the recognition program takes, the better the vision system.

At the time of this writing, the use of vision systems in robotics is advancing rapidly. Soon, they will be good enough to detect unexpected objects in the work space and prevent collisions. Programs of this sort exist now, but take so long to run that they are too slow to be useful. *The most difficult job in designing any vision system is in writing the minimum possible software.*

We have seen two examples of simple vision system programs. If the task requires, other, more complex techniques have been developed. We will examine some of these techniques in coming sections. Examples of more complex tasks include:

- distinguishing among very similar parts
- distinguishing between good and faulty parts
- recognition of parts that have a high degree of variability
- recognizing parts despite poor lighting or other sources of image degradation
- the use of color in part recognition
- recognition of parts presented to the camera in varying positions (e.g., on their sides)
- recognition of parts touching or piled onto each other
- measuring of part dimensions
- determining part orientation and distance
- determining part speed and direction of motion
- examination of parts in three dimensions

5.2 THE COMPONENTS OF A VISION SYSTEM

A complete vision system, like that shown in figure 5.3, can be considered to consist of the following components:

(A) The **image,** or **"frame."** It is affected by the light source (a major part of any vision system), the light reflectance characteristics of objects in the field of view, and how the part (if there is a part) is presented to the camera.

(B) The **camera.** CCD cameras are used almost universally now, although vidicon cameras have not disappeared, and special applications may require special cameras. The lens will be considered as part of the camera.

(C) The **image capturing system.** The circuitry and programs that periodically transfer the image from the camera to computer memory. Often called the **"framegrabber."**

(D) The **pre-processor.** Reduces the amount of information saved. May be built into the framegrabber.

(E) The **memory.** Stores the data captured for each frame. Computer random access memory (RAM) is essential, but images may also be copied to secondary memory (e.g., tape), for record keeping.

(F) The **processor.** The computer and the computer program that alters the data in memory so that the part (or its position or motion) may be recognized.

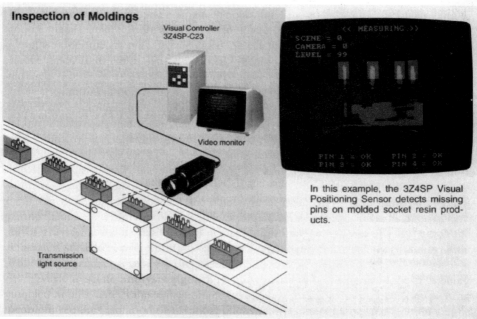

Fig. 5.3 Components of a vision system. (Photograph and diagram by permission, OMRON Canada Inc., Scarborough, Ontario, Canada.)

(G) The **recognition unit.** Where the processed image data is compared with expected results. These expected results are often called "templates." The match between the processed image and the template image is rarely perfect, so recognition results are often in the form of a "score," e.g., "The unknown part has 85% probability of being a widget." The template may be a template of a part feature instead of a template of a part, so the processor must generate an appropriate description of the image, e.g., "Has three holes." This method is useful if the vision system is required to recognize many different parts, making it impractical to provide a template for each.

(H) The **output interface,** through which the vision system outputs the results of its analysis (e.g., to a robot) or through which it outputs partially-processed data (e.g., to a video screen so an operator can monitor the system's performance).

5.2.1 The Image

When we humans think of what a vision system sees, we think in terms of the **objects** in the field of view. This perception is based on our experiences using all of our senses, including vision. The computer-based vision system, in most cases, lacks the other senses.

The image presented to a vision system's camera is **light,** nothing more. The pattern of light varies in **intensity** and **wavelength** (color) throughout the image. Light received by the camera may come from more than one light source.

The designer of the vision application must ensure that the pattern of light presented to the camera is one that can be interpreted easily. The designer can often control *what the camera sees,* ensuring that the image has minimum clutter. He or she can often control *where important features will be* in the field of view, perhaps by ensuring that workpieces are held in fixtures.

The designer can also control the pattern of light seen by the camera, by designing the light sources themselves and by blocking extraneous light (e.g., sunlight) that might affect the image. The use of carefully designed lighting is what is meant by the term **"structured lighting."** Some structured lighting techniques, shown in figure 5.4, include:

- The use of **backlit surfaces,** upon which objects rest so that silhouettes are presented to the vision system. This type of structured lighting is ideal for the type of vision system in figure 5.2.
- The use of **frontlighting from an angle,** where the shadow lengths provide depth information
- The use of light **focused to a spot,** so that the diameter of the light spot on the target can be used to determine the distance to the target from the light source
- The use of a single **line of light** (often from a laser light source), projected at the field of view at an angle. The location of the resultant light strip in the field of view is used to determine height of surfaces.
- The use of **multiple** light sources
- The use of **polarized** light
- The use of **redirected light,** to (for example) see the top and side of the same object in the same image

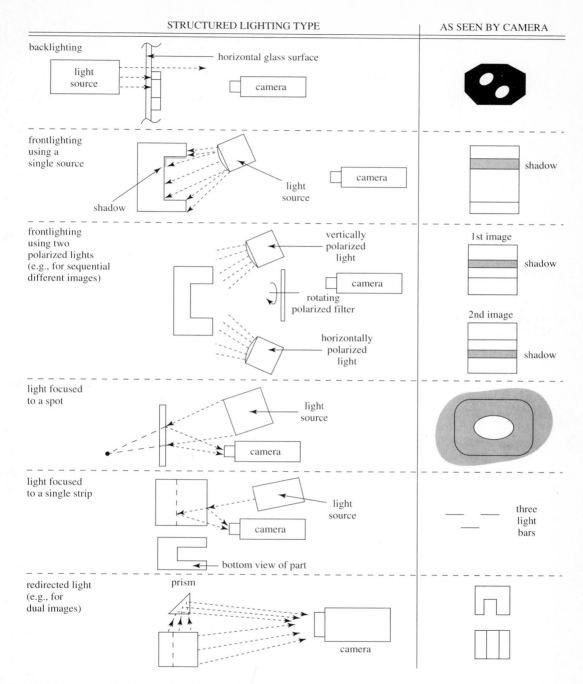

Fig. 5.4 Structured lighting techniques

- The use of different **colored** light (LEDs have narrow color bandwidths)
- The use of any combination of the above, either in a single image or in a series of images of the same scene

Sometimes, the image presented to the camera can be improved by *controlling characteristics of the objects* that will be in the field of view. Variations in the light reflectance qualities of a part, although acceptable to a human, can be disastrous in a computerized vision system.

Examples of undesirable reflectance variations include: slight color differences, differences in glossiness of painted surfaces, presence or absence of oil films, random rust patches, and even the orientation of tooling marks! Grinding mark orientations can have significant effects in a structured lighting situation. It may be better to attempt to control these characteristics of the parts than to attempt to design a vision system that will work despite such variations.

The more complex the part recognition task, the more carefully one should examine and control the image presented to the camera.

5.2.2 The Camera

Even before the light can reach the imaging surface of the camera, it must pass through the lens (or lenses) of the camera.

The Camera Lens

Perhaps *the most perfect lens of all is no lens.* In a pinhole camera (see figure 5.5), light from the field of view passes through a tiny pinhole before it hits the imaging surface in the camera. In theory, the distortion of the image should be minimal, because only the air through which the light passes can cause distortion. Unfortunately, a pinhole camera does not admit enough light to be a serious contender as a lens type.

A glass lens, as in figure 5.6, focuses light from the field of view onto an imaging surface. As light diffracts while entering and leaving the lens, and as it passes through the lens, *distortion* may occur. Solving the distortion problem requires understanding of the sources of distortion.

Diffraction occurs because light travels at different speeds through different materials. *Variations in glass composition or structure* can lead to variations in refraction angles, even inside the lens.

Imperfectly shaped lenses are another source of undesirable light diffraction. Imperfections cause more distortion if they are near the edge of the lens, because light diffracts through a larger angle there. The best portion of the image is produced by light that passes through a lens along the lens's axis. If distortion is a problem, buy a large diameter lens and use only the central portion of the lens. (Closing down the aperture blocks light from entering at the lens's edges.)

Purchased camera lens systems contain two or more lenses, matched so that imperfections in one are often corrected by the other.

Some cameras use mirrors to reflect the light to the imaging surface, avoiding lens transmissivity light distortion. These cameras are used only in highly specialized applications, such as in astronomy, and we will not discuss them here.

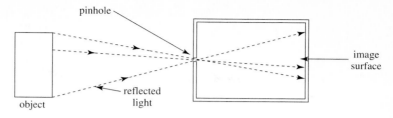

Fig. 5.5 A pinhole camera

The amount of light entering the camera can be controlled by adjusting a camera lens's aperture, or by adjusting the rate at which the computer captures images. It should not be forgotten that lens systems require careful focusing. Sensitivity to focusing error reduces as the exposure time increases and as aperture is correspondingly closed.

Image Conversion

To be useful in a computerized vision system:

- light energy must be converted to electrical energy,
- the image must be divided into discrete pixels (picture elements), and
- the brightness of the received light at each pixel must be recorded.

A color camera can be thought of as three separate cameras: one for each basic light color, each otherwise identical to a black and white camera. This discussion will therefore discuss only the ways in which a black and white camera breaks an image into pixels and how it evaluates the amount of light received at each pixel.

The two most commonly used types of camera, the **vidicon** camera and the **charge coupled device (CCD)** camera, operate quite differently from each other. We will start with the CCD camera because it has price, size, and performance characteristics that are

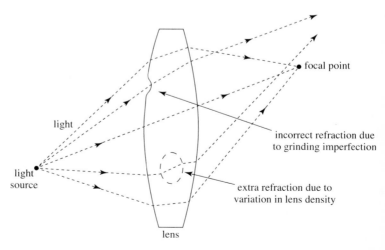

Fig. 5.6 A glass lens showing sources of distortion

rapidly leading to its dominance over vidicons. We will then discuss the vidicon because it has some characteristics that allow it to be used in applications where the data must be "pre-processed" before it is sent to the computer.

The CCD Camera. The charge coupled device is a silicon-based integrated circuit, provided to the user as a single chip. An image is focused onto the flat surface of this chip by a camera lens.

The surface of a CCD chip has columns of doped silicon, insulated from each other as shown in figure 5.7, and insulated from rows of electrodes under the silicon layer. As light strikes the "camera register" portion of the CCD surface, electrons are freed in the silicon. The electrodes under the silicon are alternately charged negative and positive, so the free electrons tend to collect close to the positive electrodes. After a period of exposure to light, the CCD will contain an electrical charge "image," in rows and columns of pixels. The charge at each pixel is proportional to the sum of the light energy received since the last time the CCD was cleared of charge.

The method of reading this charge into computer memory should be left to the section on image capturing, but in the interest of continuity, we will discuss it here. Remember that the charges are attracted to positive electrodes below the silicon. Starting from the top, the control contacts are switched in polarity one at a time. The accumulated charges are all shifted up one row. Repeated shift cycles move the whole image from the "camera" register into the "storage" register. (The move is quick, compared to the time to accumulate the charges.) After the "image" has been shifted into the "storage"

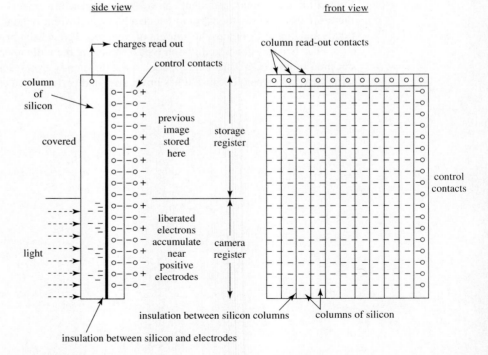

Fig. 5.7 The CCD camera

register, it is shifted out to the rest of the image capturing system, one row at a time. Meanwhile, another set of charges builds up in the "camera" register.

The CCD camera is more sensitive at low light levels than the vidicon camera, and can produce better images when there are extreme variations in brightness. On the other hand, silicon processing is still a somewhat new science, and there *tends to be a variability in light sensitivity* from pixel to pixel in a CCD. This variability can be corrected, pixel by pixel, using different amplification as each pixel is read, if the errors are "mapped" beforehand.

The Vidicon Camera. Vidicon cameras are the types of cameras used to record television programs. The operation of the camera requires many of the same techniques used in television playback technology.

In the vidicon camera, shown in figure 5.8, light (from a camera lens) passes through a thin metallic layer onto a photosensitive image surface. The light causes electrons to be released in the photosensitive layer. The negative charges tend to stay where they are created. (If very large charges build up, they will eventually spread out or "bloom.")

An electron beam sweeps across the image surface one row at a time, starting from the top row and working toward the bottom. Where the electron beam encounters zones of plentiful charge, it causes lots of charge to move into the metallic layer behind the photosensitive layer. Where it encounters zones of low charge, less charge is moved into the metallic electrode. All the charges are cleared in the process, as the electron beam passes. The charge in the electrode is thus proportional to the amount of light received at the portion of the image surface being "read" since the last time it was read.

Note that the output of the vidicon camera *does not inherently break the image into pixels* as the CCD camera did. The output of the vidicon camera looks like the time-voltage plot shown in figure 5.9.

This continuous analog output contains voltages proportional to charge levels at each horizontal sweep, and **blank sections** where the camera did not read charges because the beam was returning horizontally for another sweep, or was returning vertically for another set of sweeps.

The vidicon camera output is the type of signal that a television needs to put an image on a screen with a similar electron gun. Television signal transmissions also make

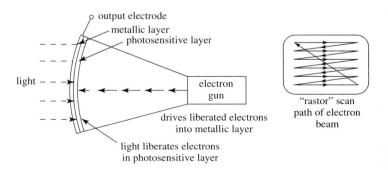

Fig. 5.8 The vidicon camera

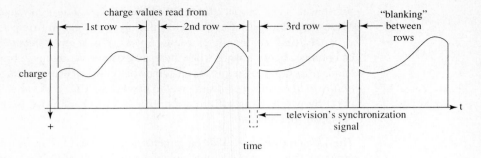

Fig. 5.9 Vidicon raster scan output

use of the blank sections, inserting a negative voltage that the television receiver looks for and uses to synchronize its playback beam location. Television's use of vidicon cameras demands that *a standard number of rows be used, and that a standard sequence of scanning of the rows* be used. The standard number of rows is different in North America (525) from that in Europe (625). Television systems use an **"interleaved"** scan sequence that starts with the top row and then sweeps each *second* row (left to right) from there to the bottom, then returns to the top and scans the missed rows. Each of the two scans of the screen takes 1/60 of a second, so the image is completely refreshed every 1/30 of a second. Interleaving of frames serves to reduce flicker and maintain more even screen brightness.

A vision system using vidicon cameras does not have to use existing television standards. Numbers of rows can be selected for the resolution required. Interleaving may be abandoned to reduce blurring due to motion. Synchronization pulses in the blank video sections are usually kept. Mass production of television technology controllers makes them cheap, so many successful image capturing systems have been built using television standards.

The *vidicon physical system does impose some constraints* on the system designer's freedom. The charge-reading process dictates that the width of each column should be roughly equal to the height of each row. The continuous analog signal must be sampled (by the image capturing system) at the correct frequency so that horizontal distance between samples is the same as vertical distance between rows. Also, the accumulated charges on the photosensitive layer tend to remain where they are created, so if horizontal sweeps are too far apart, charges between the scanned rows will be missed. On the other hand, making more horizontal sweeps takes longer, so charges may grow large between readings and "bloom" sideways, blurring the image.

5.2.3 The Image Capturing System

We have seen that cameras output a charge proportional to the quantity of light received at each segment of their imaging surface. We have also seen that the vidicon camera output signal is broken into rows but not into columns. The image capturing system, sometimes called the **"framegrabber,"** must do further modification of camera output before the resultant data can be used by a digital computer.

In some cases, the signals are "pre-processed" during image capture, to reduce the quantity of data sent to the computer. Pre-processing will be discussed in the next section. The discussion of image capturing systems in this section will assume that the whole image is captured.

The vidicon camera signal must be **sampled** and each sample must be **quantized.** The CCD camera output requires only quantization.

Sampling

Sampling of the vidicon signal breaks each row of continuous charge readings into columns of discrete charge readings. The designer of a vidicon image capturing system decides how many columns of pixels are required, then designs analog signal processing equipment that will sample each row that number of times. One type of sampling circuit would include a capacitor, and would operate as follows.

If 256 columns were required, the capacitor would be allowed to accumulate the charge from the vidicon signal for 1/256 of the horizontal sweep time. The accumulated value of the charge would then be read and converted to a digital number. The charge and read cycle would be repeated 256 times during a single horizontal sweep of the screen. This process of sampling a continuous analog signal is shown in figure 5.10. In the example, the charge level is read three times.

The single-capacitor sampling method misses much of the signal detail between sample-taking intervals. The vidicon raster scan does not stop while the capacitor is being read, so data is also lost during this part of the cycle. The use of a **multiplexer** can prevent data from being lost. Two capacitors can be used to accumulate charge from the camera output signal, with one or the other connected to the camera output at all times. The multiplexer's function would be to alternate which capacitor is connected to the Analog to Digital Converter and which is connected to the vidicon output circuit.

Taking too few samples per row can lead to types of error known as **biasing** and **aliasing.** An example of aliasing error is shown in figure 5.11, where the image is of a perforated steel part. The vidicon's output signal rises and falls at a frequency dependent on the rate at which the raster scan beam sweeps past the image's holes. If the output is sampled at a frequency only slightly greater than this frequency, the vision system would miss most of the changes, "seeing" a part with larger and fewer holes than actual.

If the sampling is done at the *same* frequency as the frequency at which the signal changes, it is entirely possible that the vision system would not "see" the part at all!

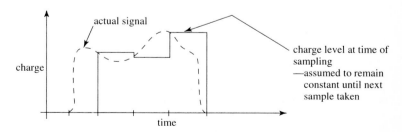

Fig. 5.10 Sampling of a continuous analog signal

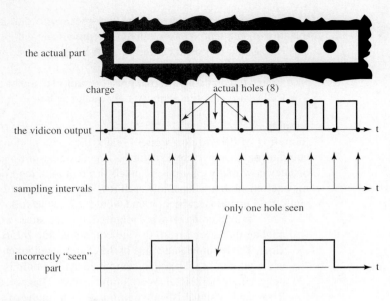

Fig. 5.11 Aliasing error

(Because each sample would be taken of a hole in the image.) This type of error, missing some feature entirely, would be an example of **biasing** error.

In general, *the frequency of the sampling should be twice the frequency at which the input signal varies,* to prevent biasing or aliasing error.

Now that we have discussed sampling of the vidicon signal, note that *the sampled vidicon signal is of the same form as the output from the CCD camera: a series of discrete charges, each of which is proportional to the amount of light received at a pixel.*

In the **CCD** camera, the number and arrangement of the pixels is fixed at the time of manufacture of the CCD. (In fact, early CCD cameras included only one row of pixels.) The CCD camera forces the user to assume that the image can be represented by points on a **square grid.** (See figure 5.12.)

In the **vidicon** camera, the number and arrangement of pixels is under the control of the system designer, at the cost of having to design and build more control circuitry. The camera allows the user to *vary the grid and pixel shape,* by varying the locations of the samples on subsequent rows (as shown in figure 5.12). If complex shapes or small features are the subject of the vision system, the center location and assumed shape of each pixel can become an important consideration.

Quantization

A digital computer cannot use analog charge values. The analog values must be converted to digital numbers. This process is known as **digitizing an image,** or **quantization.**

Quantization is performed in most vision systems by an integrated circuit chip known as an Analog to Digital Converter ("ADC" or "A/D converter"). An ADC for a vision system is chosen based on its speed and the number of bits in the digital number that it outputs. (Chapter 7 describes some types of ADCs.)

Quantization error is the result of choosing an ADC that has too low a resolution (too few bits in output binary number). Quantization error *results in a vision system that is unable to distinguish between two slightly different light levels* in the image because both light levels are assigned the same binary number. Figure 5.13 shows the results of sampling then quantizing a changing analog charge, using an imaginary two bit ADC.

The two bit ADC can output only four different binary numbers (00, 01, 10, or 11). Note that several different analog samples are assigned the same digital number. All of the digitized values are at least slightly incorrect.

Commonly available ADCs include those with outputs of 6 bits (64 possible output numbers), 8 bits (256 possible numbers), and 10 bits (1024 numbers). Since many cameras are not capable of accuracy that deserves more than six bits of resolution, using more than six bits is often unnecessary.

Speed is an important consideration in the capture of an image. The faster a single analog value can be converted to digital, the more values can be converted in the available time. This in turn means that more pixels can be captured per image, increasing resolution, and reducing the possibility of biasing or aliasing. The use of flash converters (see chapter 7) allows hundreds of values to be converted during one horizontal sweep of a raster beam in a vidicon camera.

5.2.4 The Pre-Processor

In the section on memory, we will see that the amount of data in a single image can be huge. Pre-processing is often essential *to reduce the amount of data so it can be stored.* Another major reason for pre-processing of data is *to eliminate unnecessary information so that the computer doesn't have to process it later.*

Many vision systems do pre-processing of image data before it is stored. "Framegrabbers" hardware and software packages do a significant amount of signal conversion, as we have seen, and can definitely be considered pre-processors. Most do more

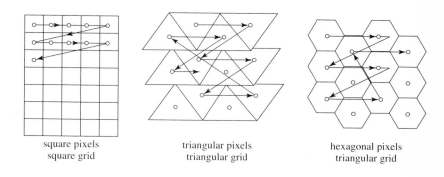

square pixels
square grid

triangular pixels
triangular grid

hexagonal pixels
triangular grid

Fig. 5.12 Assumed pixel shapes and grid shapes

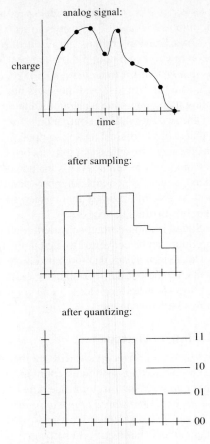

Fig. 5.13 Quantization and quantization error

than the sampling and quantization discussed in the image capturing section. These additional signal and data conversion techniques will be discussed here as "pre-processing."

Some pre-processing can be done by analog circuitry, but most is now done by digital computers. Where digital computers do pre-processing, the distinction between pre-processing and processing becomes somewhat fuzzy.

We will see that "processing" sometimes means repairing low-grade images. If repairs are necessary before a recognition program can be run, then pre-processing of uncorrected data may be undesirable! This section, therefore, assumes that the data that is being pre-processed is data from a good quality image, not badly distorted by the camera or image capturing systems. Following is a description of some pre-processing procedures used for data reduction.

Grey Scale Reduction and Thresholding

One of the techniques to reduce the total amount of data sent to the processor is to reduce the number of binary bits used to represent the light level at each pixel. This is often called *reduction in the number of grey levels.*

Two Grey Levels (Binary Images). In the extreme case, it may be possible to *reduce the number of bits to one.*

There will therefore be two possible levels of "grey." The binary number "1" might be used to depict a light pixel, so a "0" would mean a dark pixel.

Figure 5.14 shows how the computer would "see" a binary image of a dark colored telephone push-button, with the number molded in white. (In this example, the computer assumes that each pixel is square-shaped.)

This type of image representation is often called a "binary" image, not because the single digit number is any more a binary number than a standard 8 bit number, but because each *pixel* is binary (can have only two states).

Binary image representation is sufficient for recognition schemes such as those where parts can be distinguished by their total surface area. A good structured lighting technique for such a recognition system would be to backlight the field of view, so that the objects in the image would be a definite black on a definite white field. (The number on the button shown in figure 5.14, however, would not be visible unless it was frontlighted.)

Another type of processed binary image uses a "1" to represent a pixel that is different in lightness from the neighboring pixel. This image-representation would contain only the edges of the image, and is therefore called a **binary outline image.** Figure 5.15 shows how the push-button from figure 5.14 would be depicted in binary outline image form if each row of pixels is examined from left to right, one at a time, from top to bottom. This type of image representation might be useful *if the objects in the field of view can be identified based on the total lengths of vertical edges alone.* If recognition requires the total perimeter length, the processor can easily be programmed to reconstruct missing horizontal edges before running the recognition program.

The question that the vision system must face in using a binary representation of an image is: "What constitutes the dividing line between light and dark?" This dividing line is called a **"threshold"** in vision system terminology.

For a sharply backlit object, the thresholding problem is fairly simple. Figure 5.16 shows two possible results of thresholding the same backlit image. The actual edge location is shown by the dotted line. Most (square) pixels are either very dark or very light.

Fig. 5.14 A binary image

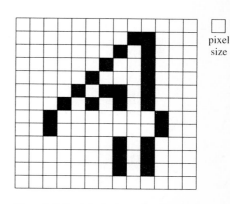

Fig. 5.15 A binary outline image

Pixels along the part edge contain both dark and light, so they are actually grey. The pre-processor must use a threshold level to assign some grey pixels as black, and others as white.

Figure 5.16 also includes a **histogram** of the darkness levels of pixels before the pre-processor assigns them all as black or white. A histogram is a count of the number of pixels at each grey level. This histogram shows two peaks, one for the large number of dark pixels, and one for the large number of light pixels.

Figure 5.16 shows how selecting threshold levels close to the histogram peaks affects the size of the object's binary image. Threshold #1 is close to the dark pixel peak, so the pre-processor will call most of the grey pixels white. The preprocessed image of the part will occupy few pixels. Threshold #2 is close to the light pixel peak, so most of the pixels along the edge will be deemed black, and therefore the preprocessed part image will be larger. As long as the chosen threshold is between these two peaks, the object will at least be visible in the binary image, but if the recognition software uses a count of the number of dark pixels to recognize the part, then a poor choice of threshold level can impair the accuracy.

The selection of a threshold value becomes more difficult (and more critical) when frontlighting is used. There may be more than two peaks in the histogram (there would be in a frontlit image of the telephone button example), reflectance qualities of otherwise identical parts might vary, and reflected light may affect the brightness of adjacent areas.

Early vision systems "solved" the threshold selection problem by requiring precise adjustment of light sources, and by requiring the user to adjust the threshold level whenever (for example) the light bulb required changing, or the characteristics of the objects in the field of view changed.

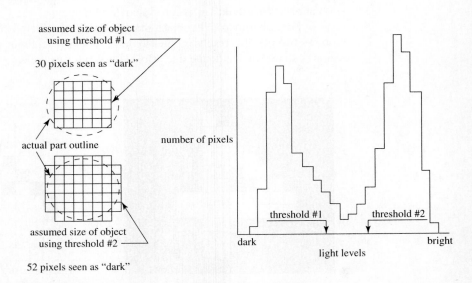

Fig. 5.16 Effect of varying threshold values in a backlit image

Newer systems can count the pixels of each lightness level in an image (and even draw the histogram). Some can select their own threshold value halfway between two peaks. Slightly more sophistication is required if there are more than two peaks.

More Than Two Grey Levels. The camera and image capturing system may be capable of more brightness range than is present or more than is useful in any one frame. Pre-processing can eliminate the unnecessary brightness levels. The binary numbering system can then be used to represent pixel brightness in the reduced brightness span with more resolution. If more resolution isn't required, data quantity can be reduced by representing pixel brightness with binary numbers with fewer bits.

Figure 5.17 shows the histogram of an image as it is outputted by an 8 bit ADC. Note that of the 256 possible grey levels, only about 50 are used. A 6 bit binary number, capable of depicting 64 grey levels, could be used. The pre-processor's job in this situation is to use the histogram to select the top and bottom of the six bit number's range. The rest of the data can be safely discarded.

If the user needs to depict a narrow band of useful brightness levels (such as that from figure 5.17) with maximum resolution (all 8 bits in this example), then the ADC's reference maximum and minimum voltage levels can be adjusted. The adjustment can be operator-controlled or computer-controlled. The new reference values should correspond to the image's useful maximum and minimum brightness levels. This method would effectively **filter** brightness levels so that any pixels with brightness above the reference maximum would be represented as maximum, and those below the minimum would be represented as minimum. Between these levels, pixels could use all of the possible binary number combinations.

Windowing

Sometimes the designer of a vision system knows that only a part of the field of view of the camera will hold useful information. If this is the case, the pre-processor can reduce the data sent to the processor by sending only data pertaining to this **region of interest** in the field of view.

In figure 5.18, the vision system's function is to measure the height of parts presented to the camera. The parts are always aligned so that the bottom edges are

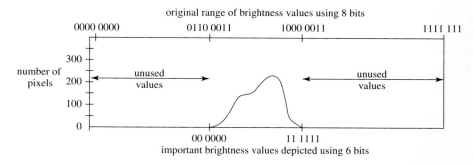

Fig. 5.17 Histogram of an 8 bit grey level image

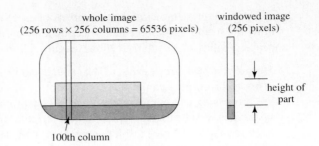

Fig. 5.18 A single column of pixels in a window to measure part width

horizontal, so the height of each object can be determined by examining a window containing a single column of pixels. The pre-processor could be programmed to capture only the 100th pixel in each row. We say that the image has been **windowed** if the pre-processor is used in this manner to blank out all the information except that in this region of interest.

Image Coding

If all the data in an image is important, yet the data representing the image must be reduced, then ways of representing the image other than pixel-by-pixel need to be found.

"Compressing" an image may be the answer, if compression methods can be used without damaging the image so that it can't be used by the image processor. Compressed images occupy less memory, and can be transmitted via data communications equipment in less time than an uncompressed image. This is important to current computer systems. It is also important in reduction of image data size for non-recognition uses of images, such as for electronic records storage and for FAX transmissions of images.

Images are often compressed and placed in individual files for storage purposes. The **file name** usually includes an extension (.TIFF, .PIFF, etc.) identifying the image formatting technique. The recently announced **FIF (Fractal Image Format)** uses fractal (or "Chaos") mathematical techniques to reduce the amount of data in an image by as much as 2500:1. The mathematical coding and decoding of an image requires an unusually long time, but FIF calculations are claimed to actually improve images!

Most image compression schemes seek to **reduce redundant information** in an image. If, for example, most of the image is a uniform grey level, a single number can be used to describe the background grey level. Only data for pixels with other grey levels needs to be saved. Run length coding can also be used to reduce redundant information.

EXAMPLE 1

In a binary image, where each pixel has a brightness of either "1" or "0," the sequence of bits describing a row of 48 pixels may look like the following, which could be stored in 48 memory locations (or in 6 eight bit memory locations):

000000000011111111111100000000111111110000000000

Run length coding would describe the sequence by replacing each of the five runs of duplicate numbers with a number describing the number of repetitions. The above sequence, run length coded, would be:

$$10, 12, 7, 9, 10$$

which, of course, must be converted to binary numbers. If 5 eight bit numbers where used:

$$00001001\ 00001010\ 00000111\ 00001000\ 00001001$$

This example demonstrates a savings of eight bits, or at least one byte of RAM memory.

EXAMPLE 2

To run length code a full row of an image with 512 columns, we would require a series of numbers, each able to depict the full width (512). The number 512 requires 9 bits to depict in binary. If the row was all zeros, the maximum savings would be achieved. The saving would be the difference, in number of bits, from 512 single bit numbers to one 9 bit number (512 − 9 = 503 bits saved). Additional bits are required to depict the start of each row, and even more bits are required if more than two grey levels are used, but a saving is often still possible. Run level coding is one of the data reduction techniques used to make FAX transmissions possible.

5.2.5 The Memory

Once an image is captured, and perhaps pre-processed, it must be stored somewhere until the main processor can use it as data. It is *usually stored in RAM*. (In some systems, the same image data is also stored on tape for a permanent record.) The unreduced data from a single image can use a lot of RAM! A 512 by 512 array of pixels, each represented by an 8 bit number, requires 256 kilobytes of memory.

$$512 \times 512 \times 8 = 2{,}097{,}152 \text{ bits}$$
$$/\ 8 = 262{,}144 \text{ bytes}$$
$$/\ 1024 = 256 \text{ kilobytes}$$

New personal computers can address a lot of memory (a 486 computer can address 4 gigabytes), and the cost of memory is falling rapidly, so memory restriction will be less of a factor in the future. On the other hand, as computers become faster and able to analyze more data in each image, image capturing devices are being designed with more pixels per image, and more grey levels per pixel, so the increased memory available in the future may not mean that more images can be stored in RAM. (A single 1024 × 768 pixel image, with 32 bit numbers describing each pixel, requires 3 megabytes of RAM!)

To compound the problem, the recognition program may dictate the need for at least *two* images in RAM: one for the image being processed, and the other for the image being captured.

Another restriction on the amount of data that can be stored in RAM has to do with *the rate at which that data can be processed*. If the images are being stored in RAM faster than they are being processed, eventually data will be lost.

Data from an image *can often be stored in memory locations in such a way as to reduce processing time and memory requirements.* If the binary word that described the grey level at a pixel is stored in a memory address that also describes the original location of the pixel, then the processor does not need to waste time interpreting the pixel's location, nor does information about the pixel's original location need to be stored as data.

Using the memory address to describe the pixel's original location is simple: If an image is made up of 256 × 256 pixels, each represented by an eight bit data word, then 65,536 eight-bit memory locations are required to store an image.

Only 16 bits of the 32 bits in a modern computer's address bus are needed to describe the memory locations required to hold these data words. The lower 8 address bits can be used to describe the column number (0 to 256), and the next lower 8 bits to describe the row number. Address location 0000 0001 0000 0000 would be used to store the binary number describing the light level of the pixel at row **1**, column **0**. Address location **1111 1111** 0000 **1111** would be used to store the binary word describing the light level of the pixel in row **255**, column **15**, etc.

The unused 16 address bits in a 32 bit computer would be used to ensure that image data is not stored in program space, or on top of other image data.

5.2.6 The Processor

The function of a vision system processor depends on what that vision system's function is. The processor may be required to:

- **reconstruct** or enhance an image
- **recognize,** classify, and/or describe the object(s) or pattern in the field of view (quality control functions would be included here)
- **analyze** the scene (for example, to enable a robot to avoid unexpected moving obstacles in its work envelope)

The simpler the processor's task can be made, the less the probability of an error. Berthold Horn, in his book *Robot Vision* (MIT Press, 1986), tells a parable of two farmers and their attempts to distinguish between two horses. After several methods have failed, they find what they believe to be a foolproof method. They measure the heights of the horses. The white one is a full two inches taller than the brown one. The point is that if you really want to make a task hard, you can find ways to avoid the easy method.

Image Enhancement

The first step in many processing tasks is to improve the image. The image as captured may contain **dimensional distortions** (perhaps caused by the imperfections in the camera lens). The image may contain **brightness distortions.** An object with a uniform white surface, for example, may be perceived as having several grey levels if some areas of the object are illuminated more brightly than others.

Very bright areas of the image may have caused very high charge levels in the image capturing device. These charges may have bled into neighboring portions of the imaging surface, so that these neighboring pixels seem brighter than they actually are. This is known as **"blooming."**

"Ghosting" of previous images may be present if the camera didn't completely neutralize the charges during the capture of that previous image.

If these problems exist, then the image may have to be corrected, or "reconstructed," before it can be used as data.

The image may be deficient for other reasons. It may contain **pixels** that are so *large* that determining the actual locations of edges of surfaces is difficult. It may contain pixels in which the analog to digital conversion has produced incorrect results **("noise").** The data may have been **damaged** during transmission before or after analog to digital conversion.

The presence of a few incorrect pixels may have resulted in the pre-processor incorrectly selecting maximum and minimum levels for the grey scale range, so that the true image is **compressed** into only a few levels of grey. If the time taken to capture the image from the camera's image surface was long relative to the time required for objects in the image to move, then **blurring** may be present. Finally, the camera might actually have been **poorly focused!** In these cases, image enhancement may be necessary before the image data is usable. In some cases, the reconstruction may be simple.

If, for example, the distortion caused by a camera lens can be mapped, then appropriate correction factors can be applied to the portions of the image known to be affected. This method, known as applying the **"inverse transform,"** can be applied wherever the cause of the distorted image is known and is constant. It can also be used to correct for variations in the light levels illuminating the field of view. The drawback is that the data enhancement program may have to contain several correction factors for each pixel.

The use of **"Gaussian filters"** reduces the need for the storage of correction factors. If, for example, it is known that there is a distortion centered on a certain pixel in the image, then it can often be assumed that the neighboring pixels are also distorted, but that the amount of distortion is less for the more remote pixels. The reverse transform would have a "normal" or "Gaussian" distribution about the center. This statistically-based method of determining correction factors requires that only data for the mean, standard deviation, and center of each distortion source needs to be stored. Figure 5.19 shows how Gaussian filters would be used to correct for two overlapping distortions in an image.

Noise can sometimes be filtered out of a binary image by comparing adjacent pixels' brightness values, and then eliminating or reducing suspiciously large differences in brightness levels.

Any program that changes the data that is received from the image capturing system runs two types of risk:

- **"Type 1" error** occurs when faulty data is not corrected.
- **"Type 2" error** occurs when good data is assumed faulty, and then is changed!

The system designer must find the correct balance of potential error, then implement the appropriate image enhancement scheme.

As a final reminder, remember the farmers and their horses. If a way can be found to use the vision system without image enhancement, then that method should be seriously considered. While image enhancement may reduce the possibility of Type 1 error in subsequent image recognition (an object not being recognized), it also increases the possibility of Type 2 error (a recognizable object not being recognized because its image has been altered during "enhancement").

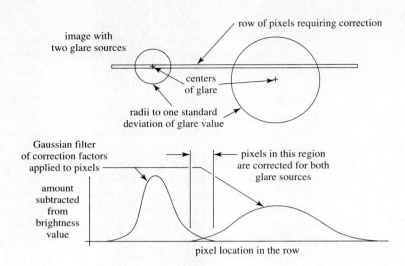

Fig. 5.19 Gaussian filter use in correcting images

Object or Pattern Recognition

"Recognition" of objects or patterns in an image forms the bulk of current vision system applications. Vision systems are being used in industry to examine unknown objects presented to the camera, and to make decisions about the objects. The decisions may be:

- *what* the object is (e.g., so that an automation system can direct it to the appropriate chute)
- the *location* of the object, and what *motion* it has (e.g., to enable a robot to pick it up)
- the *orientation* of the object (e.g., so that a robot can be instructed about required gripper orientation)
- *other* evaluations of part dimensioning, completeness, color, etc. (e.g., so that faulty parts can be rejected before automatically being assembled with other parts, or so that statistical process control of the manufacturing equipment can be implemented)

If recognition of only a few parts (or features of a part) is all that is required of the vision system, and if the characteristics of the parts are known to the designer of the vision system, the parts can be recognized by comparing the part in the image to a **template.** Because of memory limitations, and because a separate template is required for each possible part, the "template" is not usually a very complete template.

One or Two Dimensions. In the simplest case, a template might be a *single number,* perhaps representing the width (one dimensional data), or the surface area of the part (two dimensional data) in terms of the number of pixels the image of that part would fill. In other cases, the template might consist of *several different multi-dimensional* *characteristics* of the part. Typical part or feature descriptors that might be used for a template are shown in figure 5.20 (pp. 134–5). After examining figure 5.20, the reader who wishes to learn more about vision systems, but who isn't interested in the specific techniques of data processing, can skip ahead to the section discussing three dimensions.

In some of the examples of templates shown in figure 5.20, it is sufficient for the recognition program to find objects by examining the pixels in an image, one row at a time, from left to right, and from top to bottom. If a pixel's light level is very different from its neighboring pixel's light level, then an edge of an object has been found. This method, known as the **"local counting"** method, is often adequate where the image contains only a few simple objects.

Local counting techniques fail in situations like that demonstrated in figure 5.21 (a), p. 136. The system thinks that it is counting pixels for two distinct objects until midway through the fifth horizontal scan, where a pixel is found that touches both "objects." Which region is this pixel a part of? If the problem in figure 5.21 (a) is "solved" by assuming that all pixels with a common grey level are parts of the same object, then we create a new set of problems. Figure 5.21 (b) shows several small parts that might be "recognized" as a single large part because they have the same grey level. It is necessary to evaluate the image in such a way that the pixels representing each separate part are counted separately.

The method known as **region growing** works well to solve both of the above problems. As shown in figure 5.22, p. 136, the region growing technique works as follows:

1. select any pixel (possibly at row 1, column 1), and assign that pixel the value "one."

2. evaluate each neighboring pixel to find if it has the same grey level. Each pixel with the same grey level is assigned the same value.

3. select one of the neighboring pixels with the same grey level. Go to step 2. If there are no more pixels of the same grey level next to this region, go to step 4.

4. select a pixel that hasn't been counted yet (possibly an adjacent pixel with a different grey level). Assign it the next sequential value ("two," "three," etc.).

5. go to step 2.

With the region growing technique, eventually each separate region of each grey level will be assigned a unique identifying value, and the recognition system can proceed with the template-matching routines.

Figure 5.23, p. 137, demonstrates **the concept of connectedness.** To be part of a region, it is necessary that a pixel be adjacent, or "connected," to more than one pixel in the region. For *maximum* confidence that a pixel is within a region that extends in two directions from that pixel, it is necessary that the pixel has a connectedness of 6 (touches six pixels of the same region).

In some cases, an **"iterative modification"** technique may be used to reduce the data in an image, while the image is being processed for recognition. In the iterative modification technique, *regions of identical grey levels are reduced in size* until each region is only a skeleton of its original size. With each iteration, pixels with a connectedness of 6 or greater are eliminated.

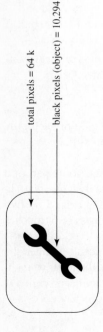

total pixels = 64 k
black pixels (object) = 10,294

(a) the surface area, in terms of numbers of pixels

three gray levels:
0000 level (background) = 40 k pixels
1010 level (dark plastic) = 20 k pixels
0110 level (light plastic) = 4 k pixels

total object = 24 k pixels, of which the number occupies:
4 k/24 k = 17% of surface area

(b) the number of pixels with specific gray levels, perhaps divided by the total number of pixels in the complete object

there would be 2 gray levels in example (b), after discarding the background gray level:
—minimum brightness is 1010 level (dark plastic)
—maximum brightness is 0110 level (light plastic)
the average (binary) gray level is:
(20 k*1010 + 4 k*0110) / 24 k = 1001

(c) the maximum, minimum, or average gray level present

5000 pixels on perimeter

(d) the length of the perimeter of the object

150 pixels at maximum width

(e) the maximum width (or height) of the object

250 pixels in longest uninterrupted row
150 pixels in longest uninterrupted column
ratio = 250 / 150 = 1.7

(f) the width-to-height ratio

200 pixels in longest uninterrupted line in any direction
—longest line is at 40 degrees

(g) the length or orientation of the maximum length line bisecting the part

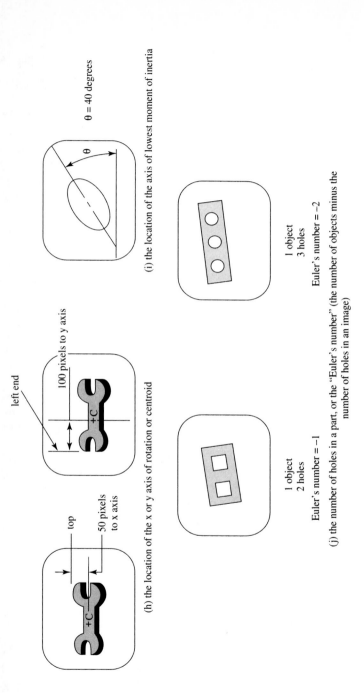

Fig. 5.20 Template features.

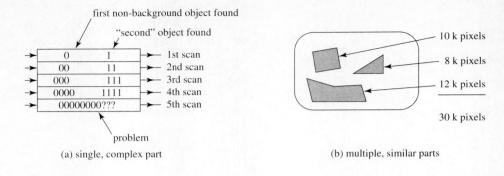

(a) single, complex part

(b) multiple, similar parts

Fig. 5.21 Problems encountered with local counting methods: (a) single, complex part; (b) multiple, similar parts

After iterative modification, all the original regions still exist (although they are much smaller) and their positions relative to other regions will not have changed. The number of iterations required to reduce each region to one pixel can be used to describe the region's original size. It is possible to reduce the image further by shrinking the regions to nothing, and counting the number of grey levels that are "destroyed" in the process.

Three Dimensions. To recognize a part or a feature of a part, it may be necessary to analyze two-dimensional data from the captured image so that information about the actual three-dimensional object can be extracted.

Structured light technique can be used to allow the processor to *estimate distances from the camera to a surface* on the part in the field of view. One method, shown in figure 5.24, is to use an angled line of light, often laser generated, to illuminate the objects in the field of view. Trigonometric functions in the recognition program can then be used to calculate the distance of each surface from the camera using the location of the observed reflected light. If the column of light sweeps across the whole field of view, the distance to each surface (except surfaces in the shadow of other surfaces) can be calculated.

The *orientation of a surface dictates how much light that surface will reflect* into the camera. Figure 5.25, p. 138, demonstrates that the maximum light is received at the camera if the surface is at a right angle to a line exactly bisecting the angle between the light source and the camera.

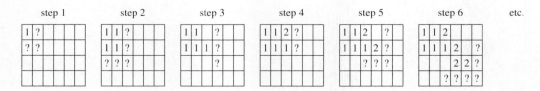

Fig. 5.22 Region growing

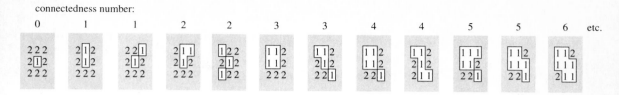

Fig. 5.23 Connectedness

If the surface inclines *in any direction* away from the bisecting line, then the amount of light reflected into the camera is reduced. If the recognition system program includes an experimentally-determined lookup table, it can use the brightness data from each pixel to determine the angle of the surface relative to this bisecting line. Such a lookup table is called a **"reflectance map."**

Determining which direction the surface inclines is another problem. One method of determining the orientation of a surface angle is to use two cameras, but it is possible to use a single camera to determine surface orientation. If light comes from directly behind a camera, the single image can be used to estimate three-dimensional information about a surface.

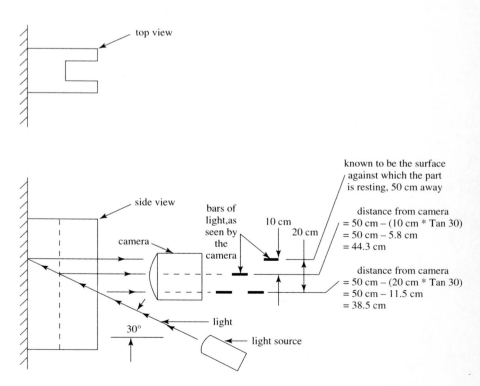

Fig. 5.24 Columnated laser light to detect distances

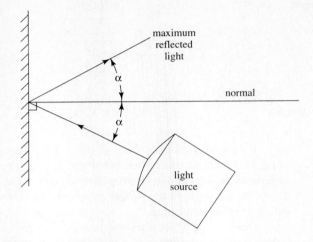

Fig. 5.25 Influence of surface angle on reflectance

The brightest portion of the image is assumed to be the surface closest to the camera and perpendicular to the camera's axis. Each adjacent pixel with a lower light reflectance is then assumed to incline away from perpendicular. Figure 5.26 illustrates. The normal of each pixel in the image is assumed to intersect the lens axis line either below or above the surfaces. This method is still not as good as using two images of the same scene.

Once the orientations of the surfaces are determined, it is possible to "correct" for the true area of each surface, using simple trigonometry. Figure 5.27 demonstrates the trigonometric process that can yield the actual length of surfaces, when you know

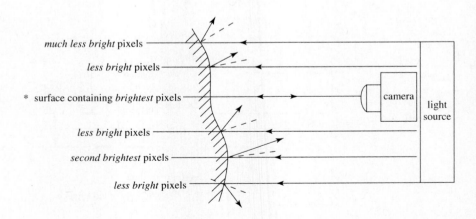

Fig. 5.26 Estimating surface orientation from a single image

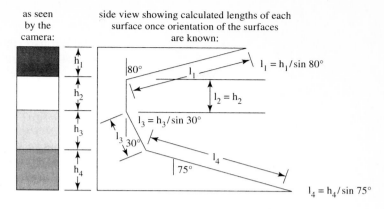

Fig. 5.27 Using orientation data to correct for surface area

their height in the image and their orientation. The distance from the camera to each pixel in the image can then be determined by integrating (accumulating) the resultant depth from each area.

When the size and orientation have been determined for each surface, by any method, the image can be stored as a **needle diagram.** In a needle diagram, each surface in an image is described as a vector (two numbers: a length number and an orientation number). The *length* number indicates the area of the surface, and the *orientation* number the orientation of the surface. The graphical version of a needle diagram is shown in figure 5.28.

Templates of three-dimensional images can be complex. It is often desirable to describe a three-dimensional image as a **tessellated sphere** against which a tessellated sphere template can be compared.

The concept of a tessellated sphere is as follows: The infinite number of orientations in three dimensions is reduced to a manageable number of orientations, just as a geodesic dome is made up of a manageable number of flat surfaces (tessellations) that approximate a sphere. Memory space is set aside for one number for each "tessellation."

Non-spherical shapes in the real world might contain large surface areas with one of the sphere's tessellation orientations, but no surface area at another orientation. When

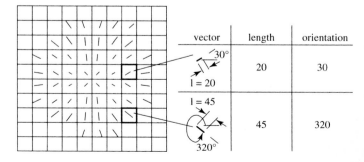

Fig. 5.28 A needle diagram of a sphere

real-world shapes are represented by tessellated spheres, the total amount of surface area for each surface orientation is represented by the number stored for that tessellation. Shown graphically, the surface area numbers protrude from the appropriate tessellation.

As shown in figure 5.29, a cube would be represented by a tessellated sphere with only three "needles" with a length greater than zero. (Three faces of the cube are hidden, as is the back surface of the sphere.) Such a tessellated sphere with attached vectors is called a **"spike model."**

Scene Analysis

The human being, with two eyes, a remarkably adaptive brain, and the ability to move and to change the light illuminating a field of view, can analyze scenes and *reach conclusions about them without being specifically "programmed" for each situation.* Vision systems are also being given this ability, although the current level of sophistication is still crude.

It is now economically feasible in some applications to develop and install scene analysis types of vision systems. Where robots work in environments that change unpredictably, for example, it is necessary for them to "see" the environment, to make decisions about tasks to be performed, and indeed, whether it is safe to do anything. In satellite probes designed to land on other planets or to fly past comets, communication with home base becomes subject to message transmission lags, so it is necessary that the on-board computer can evaluate and respond to unpredictable visible situations. Where events occur too fast for a human operator, as in a modern fighter plane, it is necessary to equip the plane itself with "eyes" and the intelligence to respond to sensed changes in the environment. This section will deal with current issues in scene analysis. The reader not interested in details can skip ahead to the section discussing recognition units.

Scene analysis is required so that *a complex image can be broken down into recognizable features,* or so that other *essential information can be obtained from the image or sequence of images.* An example of retrieving features from complex images would be where one part would have to be identified in a pile of identical parts. An example of extracting "other information" from images is in guidance systems, where the speed of objects must be calculated.

The Bin Picking Problem. If a robot has to pick a part out of a bin of jumbled parts, it must be provided with information about parts that are accessible. If a vision system has to provide this information, it must analyze the whole image so that it can:

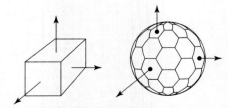

only three faces visible so only three tessellations have vectors

Fig. 5.29 The tessellated sphere (spike model) to depict three dimensional objects

- differentiate between the individual parts
- recognize each different part
- recognize the *orientation and location* in space of each part
- determine which of the parts are accessible to the robot

Two-dimensional solutions exist, but we will examine the problem, and its solution, in three dimensions.

To evaluate three-dimensional images, *it is ideal to have three cameras (or a single camera that can be moved to three locations).*

From the three images, a three-dimensional model of the surface of the jumble of parts can be generated, and the data may be used in "finding" individual parts. There will be portions of the image that cannot be evaluated because the light levels at those points are not acceptable light levels. This may be because a portion of the image is in the shadow of another part, or may be because light is reflecting from an adjacent part. Partial shadows and reflectance will result in distortions of the shape of the three dimensional model. The vision system *then must select a point in the image,* and determine the limits of the object containing that point.

A logical starting point would be at the highest point in the three-dimensional image. Otherwise, any bright point may be chosen, on the assumption that a high light level means the part must be close to the camera. From this single pixel, adjacent pixels are examined. If the light level is similar at these pixels, they are assumed to be part of the same surface of the same part. Data from these pixels can be saved for later recognition schemes.

If the light level is drastically different at an adjacent pixel, there are three possible causes:

- if the light level is significantly lower, *an edge of the part may have been encountered.* A very low light level might mean a sharp edge, with shadow beside it. An intermediate light level may mean a rounded edge.
- if the light level is significantly lower, *another part may be on top of this part,* casting a shadow.
- if the light level is significantly higher, *light may be reflecting* off an adjacent part. The adjacent part may or may not be an obstruction to the robot if the robot tries to pick the part up.

To overcome some of the above problems, it may be advisable to **grow regions.** Assume that adjacent pixels are portions of the same part if there are no sharp differences in light level. The "individual part" regions being grown may be limited to include only surfaces that incline at 45 degrees or less from the camera. In this way, ambiguous regions of each part may be ignored altogether.

When all the distinct regions have been "grown," each can be compared against the template. The region that most closely matches the single template can then be assumed to be the part that is on top. Any regions that do not match the template, or match poorly, can be assumed to be partly underneath other parts.

If the parts in the bin are not all the same, a similar technique can be used, but each grown region must be compared with all templates. Again, the best match would be assumed to indicate the topmost part.

Movement in the Field of View. In situations such as where a vision system assists a robot to pick up objects from moving conveyor belts, or where a vision system assists the pilot of a fighter plane to shoot down incoming missiles, the vision system must *track and evaluate the trajectories of objects* in the field of view.

Different approaches to the analysis of scenes for determining velocities and accelerations of objects can be taken. The choice *depends on how objects are expected to move.* Figure 5.30 shows that an object can have a translational movement along any of three axes, and may have rotational motion about any of these axes.

If *rotation is not present, the velocity of the part can be determined by identifying a single point (called a "fiducial") in sequential images,* and determining the change in its position per change in time between images. Acceleration is determined by finding the change in velocity per change in time. In fact, simply identifying the same point in sequential images may be a difficult task, if one remembers that the reflectance of individual points changes as the points move past a light source.

If the part has a motion component along the z-axis (normal to the camera's lens), then the *position may be evaluated by using the square law of brightness versus distance from the observer,* or perhaps by observing the change in size of the object.

Motion that includes rotation may be miscalculated or may not be detected at all if only a single point on the object is observed. In fact, it has been estimated that, *for complete evaluation of motion of an object, seven points must be observed.*

It may be a good idea in some cases to *use a needle diagram to describe the motion of each point* in the field of view. This time, the needles depict the direction and velocity (or acceleration) of each point in the image. Figure 5.31 is a needle diagram depicting the velocity of rotation of the Earth.

5.2.7 The Recognition Unit

Once the data in the image has been corrected or enhanced, then processed, it can be compared against the template (if a template is available) and the recognition routine can proceed.

It is rare for an image of an object to match a template exactly. If, for example, the template is a count of the number of pixels that an object should occupy, then the threshold(s) used to convert analog light levels to binary numbers may affect the count (remember figure 5.16). In fact, if the resolution of the image capturing system is great enough, actual variations in the true size of the objects may mean that none of the objects will match their fixed templates exactly.

linear motion along x axis
linear motion along y axis
linear motion along z axis

rotary motion about x axis
rotary motion about y axis
rotary motion about z axis

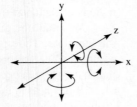

Fig. 5.30 Six degrees of freedom

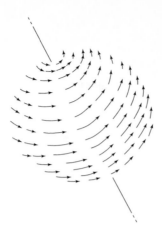

Fig. 5.31 Velocity needle diagram of the Earth

Recognition of an object or other feature should be handled as a statistical probability. The closer the processed data matches the template, the more probable it is that the object is in fact the object represented by that template. If a template describes the size of an object as, say, 3000 pixels, then an object occupying 3000 pixels might earn a probability of 1.00 of being the object described by the template, an object occupying 2985 pixels might score a probability of only .75, and a 2900 pixel object might score only .10.

If there is more than one template, then the data for the object in the camera's field of view may have to be scored against each template. The match with the highest score identifies the object, providing that score is high enough, of course.

There *must be a cutoff point, in terms of the probability,* within which the vision system can safely claim to have recognized the object. If the consequences of a **Type 2 error** (recognizing an object as something it is not) are serious, then a higher probability must be obtained before a recognition can be claimed. If a vision system reads postal codes from envelopes before the letters are automatically directed to destinations within the country, then the delivery delay that would be caused by sending a letter 2000 miles out of the way may cause the programmer of the vision system to require a .98 probability before the system can act on a "successful" reading of the code.

A fighter plane vision system, on the other hand, cannot afford to make a **Type 1 error** (failure to recognize an object) in recognizing enemy missiles. Such a recognition routine might be prepared to accept a probability of .05 as sufficient to respond by shooting down the object in the field of view. (Let passenger planes and birds in the area look out for themselves!)

Sometimes, the probability required before a recognition is accepted can be adjusted by other circumstances. In the fighter plane example, the vision system should be programmed to require a higher probability for recognizing a missile if the plane is flying in formation with other planes from the same squadron. In the postal code example, certain postal codes (e.g., those for locations in New York City) may in fact be more frequent, and thus require lower probability than others.

The Source of the Template

Where does the template come from? For simple templates, it may be easier, and therefore cheaper, for *the programmer to include them* in the recognition program as it is being written. A more flexible approach is to write the program so that *data for templates can be entered later.* The user of the vision system can then adjust the template to compensate for external factors (e.g., surface area of a part may change with minor design changes). Such an approach may allow the addition of more templates as the variety of parts changes.

The most flexible approach is where the supplier of "generic" vision systems writes the recognition schemes to use a very strong image processing method and a complex template model so that the program does not need to be rewritten for each application. *The template may then be "taught" to the vision system in teach mode.*

In **teach mode,** several images are presented to the camera, one at a time. The vision system processes each image, generating data that would normally be compared against a template. The user then tells the vision system (through a keyboard) whether this image should match the template or not. The vision system uses the processed image data, and the operator's feedback, to generate its own template.

5.2.8 The Output Interface

A vision system is useful only if adequate output can be obtained from it. The most commonly-used form of data output is via an **RS 232 serial port** (see chapter 7).

The amount of data outputted by the vision system depends on the amount of data required by the application. In the simplest case, the vision system might indicate to a robot only that a *workpiece has arrived.* One parallel port bit would be sufficient.

- More complete data output might be required if the vision system must provide the robot with *positional data,* or with data sufficient for a robot to grasp a moving object. A series of asynchronous RS 232 serial transmissions would be adequate.
- Data may have to be outputted by the vision system *to allow records to be kept.* Here, synchronous serial data output to a tape recorder may be sufficient. The data stored may be the raw data or the fully processed data.

It may be necessary for the complete unprocessed image data or the processed image data to be *outputted to a monitor* so that a human operator can monitor the quality of the processing.

Sometimes, the output may have to be *outputted to more than one output device,* for example, to a robot and to a computer running a statistical quality control program. A local area network (LAN) may be used. The synchronous serial data packets must be prefixed by addresses identifying the devices for which each is intended.

5.3 SUMMARY

A vision system requires a well defined **image,** or **pattern of light. Structured lighting** can improve vision system performance. Images may be distorted by **lens** imperfections, or may be perceived as distorted due to unevenness in the light distribution or reflection characteristics.

Light, captured by a **camera,** must be **converted to voltage levels** per pixel. The CCD camera inherently does the breaking up of an image into **pixels,** whereas a vidicon's output must be sampled. Both cameras inherently convert light to voltage. The resultant voltage levels must be **digitized** by analog to digital converters for later processing. Sampling and digitizing can both lead to loss of detail in an image.

Pre-processing of the data during or after image capturing can make data storage needs less, and data processing time shorter. Pre-processing techniques include: **windowing** to reduce the number of pixels, reduction in the number of **grey levels** (and thus the size of binary numbers) required to describe the important features in the image, and the use of **data compression** techniques. Storage of binary numbers representing light levels at individual pixels can occur in memory locations that have addresses describing the actual pixel location, eliminating the need to store location data.

Images may or may not need to be **enhanced** before later processing and recognition. If they are to be enhanced to reduce the probability of making **Type 1 errors (non-recognition),** then **Type 2 errors (mis-recognition)** may become more likely. Enhancement techniques can include a pixel-by-pixel reverse transform to reverse known error sources, or may require more complex algorithms to recognize and remove improbable data.

Before recognition, processing techniques can be used to **reduce the image data** to a more manageable, if more abstract, form. Pixels of a given light level can be counted. Lengths of perimeters may be calculated. Selected dimensions, or locations of features, can be calculated. Holes can be counted. Techniques such as region growing may be required.

Evaluation of depth (the third dimension) in a two-dimensional image can be performed by assuming a nearest point, recognizing different grey levels as different orientations, then trigonometrically calculating depths.

Three-dimensional images can be standardized by developing **spike models** of the image, in which a vector's ("spike's") length serves to depict the total area of surface with a given orientation, and where a "spike" is calculated for each of the limited number of possible orientations.

Once the data is processed, the objects in an image may be **recognized** by comparing them with **templates** of the same form (number of pixels in area, location of center of gravity, spike model, etc.). Since captured images will not exactly match the templates, **statistical probability routines** are used to maximize chances of recognition while minimizing chances of misrecognition.

Various **scene analysis** algorithms allow parts to be recognized, despite being jumbled in bins, and evaluate the complex motions of moving objects. The **bin-picking problem** requires isolation of the most unobstructed object before recognition can be successful. Several parts may have to be isolated, by three-dimensional region growing, and compared with templates before a successful recognition. Evaluation of simple translational **motion** requires recognition of the same point in separately captured images. More complex motion of a target can be evaluated only after identifying the same points in several images, and using position changes to calculate trajectory, velocity, and acceleration. Velocity needle diagrams, similar to spike diagrams, can be used to describe the velocity of several points.

Most vision systems provide **output** to robots, monitors, mass storage devices, etc., via serial interfaces such as an RS 232 interface.

REVIEW QUESTIONS

1. What does the camera receive from the target?
2. What is the image on the camera's imaging surface actually made up of before it is read by the "framegrabber"?
3. What is a pixel?
4. Describe the steps in reading the image from a CCD camera into computer memory. Assume no data reduction techniques are required and that each pixel is described with an 8 bit data word.
5. If a vidicon camera with a square imaging surface is programmed to scan 512 lines as it reads the image, describe the steps in processing its analog output to read the image into computer memory. Assume no data reduction is required, and that each pixel is described by an 8 bit data word.
6. How many kilobytes of memory are required for the image captured through the vidicon camera in question 5? You will have to select and specify a pixel shape before answering. Assume each pixel requires an 8 bit data word.
7. What is biasing error? How would you correct this type of error?
8. What is quantization error? How would you correct the problem if quantization error was significant?
9. What is a binary image and how would adjusting of threshold values affect the size of an object in a binary image?
10. What is a light-level histogram of pixels? How could it be used to reduce memory requirements without losing image data?
11. What is meant when it is said that an image is corrected by a reverse transformation? What is required before a reverse transform correction can be considered?
12. What is a template, as used in vision system recognition schemes?
13. What is a type 1 error in image recognition? Type 2?

14. What is meant by "feature recognition" as the term applies to processing a simple two-dimensional image for recognition? What might the template consist of?
15. What is region growing and why is it sometimes necessary in the processing of a frontlight image?
16. Three-dimensional images can be processed until the data describes a tessellated sphere. What is a tessellation? What does the needle length for each tessellation mean?
17. After a vision system has processed data from a pile of parts, it compares each of the "regions" of the image with the template it has of the single type of part in the bin. Why might it not recognize some regions?
18. Identify the 6 degrees of freedom of motion.
19. A vision system monitors parts arriving on a conveyor belt, so that a robot can pick them up. There is only one type of part, but the parts are simply thrown onto the belt. What information must the vision system acquire from the image to give to the robot?
20. Describe the function of each part of a vision system that must differentiate between a dinner fork and a spoon lying side by side (not touching). Do not neglect the image and the vision system output. You must use a CCD camera, but all other choices are for you to make.

PART C

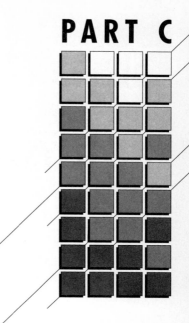

AUTOMATION CONTROLS

6	SERVOSYSTEMS	150
7	CONTROLLERS: DIGITAL COMPUTERS AND ANALOG DEVICES	192

CHAPTER 6

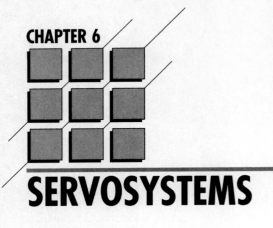

SERVOSYSTEMS

CHAPTER OBJECTIVES

After completing this chapter, the reader will be able to:

- Describe the difference between open loop and closed loop control.
- Interpret simple transfer function and graphical description of servosystem response.
- Discuss the characteristics and reasons for unstable response.
- Improve the response of servocontrolled systems by modifying:
  - error signal shape
  - physical system characteristics
  - selection and tuning of the controller
- Correctly use Proportional Integral Derivative (PID) control.
- Select the most appropriate control configuration from the following:
  - single or nested loop
  - feedback or feedforward control
  - true or synthetic feedback
- Discuss developments and the current state-of-the-art in digital controllers for servosystems.

6.0 SERVOSYSTEMS

In the Introduction to Automation chapter we saw that Computer Integrated Manufacturing requires automated control. We saw that automated control requires a control system to sense a system's output and to drive actuators until the output is correct. The sensor-controller-actuator configuration is called "closed loop control," or "process control," or **"servocontrol."**

Today, almost every manufacturer uses automated processes. Specialty equipment suppliers and even PLC suppliers offer affordable, off-the-shelf, servosystem controllers called "PID controllers." New engineers and technologists must be able to design and operate servocontrol systems.

6.1 CLOSED LOOP CONTROL SYSTEMS: A SIMPLE BLOCK DIAGRAM AND EXAMPLE EXPLAINED

In chapter 1, we discussed open versus closed loop control systems. Figure 6.1 shows basic open and closed loop systems. In the **open loop** system, the operator expects that the system output (velocity) will be proportionally related to the input signal (pedal position). If it isn't, then changing the input signal is the only way to get the required output.

A **closed loop control** system, such as the one in figure 6.1, alters its own output to meet the requirement demanded by the input signal (thermostat setting). The actual system output (room temperature) is measured by a sensor. Sensory feedback is then compared with the input signal and the actuators' output is modified (furnace turned on or off) until the output is correct.

Figure 6.2 shows a typical block diagram of a simple servosystem. It shows that a **setpoint** (e_{in}) must be provided. This setpoint may be provided manually, or may be supplied by a programmable device such as a robot controller.

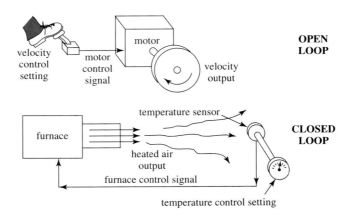

Fig. 6.1 Open and Closed loop control systems

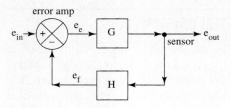

Fig. 6.2 A typical servosystem block diagram

The setpoint signal passes through an **error amplifier**, where it is reduced or increased by a feedback signal, resulting in an **error signal** (e_e) that is proportional to the difference between the setpoint and the feedback. In the temperature control example, the error signal indicates only "warm enough" or "not warm enough." Other control systems generate error signals proportional to the amount of error.

The error signal is responded to by an **actuator** (G) that does some work in response to the signal received. The actuator may in fact be several amplifiers and actuators in series. (G) also describes effects of the environment.

A **sensor** measures the **output of the system** (e_{out}), providing a feedback signal proportional to the actual system output. The output can be changed by the servosystem's environment as well as by the actuator. Environmental conditions that affect system output are called **disturbances.** The sensor's output signal may require adjusting, or "conditioning," so that it will be compatible with the original setpoint signal. The sensor and any further signal conditioning are described by H.

The conditioned **feedback signal** (e_f) is subtracted from the setpoint signal (e_{in}) in the error amplifier, resulting in a new error signal (e_e), which is again supplied to the amplifier. As long as there is a difference between the setpoint and the sensed output, the actuator will receive a signal to correct the difference.

Figure 6.3 shows an actual servosystem, like that depicted in the block diagram we have just examined. This actual servosystem controls the position of one axis of a machine bed. The servosystem is under the control of an NC milling machine controller (computer), which has been programmed to move the machine bed (and therefore the workpiece) through a series of positions past a stationary rotating milling cutter.

The NC mill's setpoint is provided by the computer as a digital number, which is converted to a DC voltage by a digital to analog converter. The DC voltage (e_{in}) remains constant until the NC program changes it. In the example, the possible DC setpoint values can range from 0 to 10 volts, which are interpreted by the servosystem as a desired position for the machine bed. (0 volts means move the bed to the 0 inch position, 10 volts means the 20 inch position.)

The actual position of the machine bed is indicated by the feedback voltage to the error amplifier, and is subtracted from the setpoint voltage. For example, the setpoint voltage might be 5 volts (meaning the desired bed position is 10 inches), while the feedback signal is 0 volts (because the bed is currently at 0 inches), so the error amplifier output is:

$$\begin{aligned} e_e &= e_{in} - e_f \\ &= 5V - 0V \\ &= 5 \text{ volts (for 10 inches error)} \end{aligned}$$

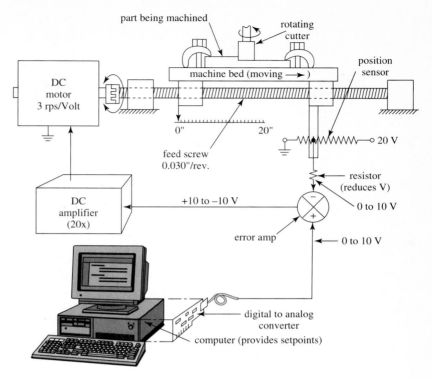

Fig. 6.3 An NC controlled, DC motor driven, position servosystem

In our NC system, the "actuator" block from the block diagram includes an amplifier, a DC motor, and a screw feed. The low current, 5 volt signal isn't sufficient to drive an average motor, so it must be amplified first. The motor converts current into torque and therefore into rotary motion. The screw feed converts rotary motion into linear motion. The total gain (G) of this series of components is:

$$G_{total} = G_{amp.} * G_{motor} * G_{screw}$$

so, if:

- the gain of the electric amplifier is 20 volts output/volt, and,
- the gain of the motor is 3 revolutions per second(rps)/volt, and,
- the gain of the screw is 0.030 linear inches/revolution,

then the total gain of the three components is:

$$G_{total} = 20 \text{ volts/volt} * 3 \text{ rps/volt} * 0.030 \text{ in./rev.}$$
$$= 60 \text{ rps/volt} * 0.030 \text{ in./rev}$$
$$= 1.8 \text{ inches per second/volt}$$

So, since the error signal that this series of components is receiving is 5 volts, the output of the "amplifier" (assuming instantaneous acceleration) is a velocity of 9.0 inches per second. At this rate, the machine bed would move to its new position in just over a second, but let's look at the rest of the system before we jump to conclusions!

At 9 inches per second, after 0.1 seconds the machine bed will have moved 0.9 inches closer to its demanded position. The sensor provides a voltage feedback proportional to the position of the machine bed. Specifically, it outputs 1 volt per inch. The sensor's output is "conditioned" by the signal conditioning equipment, which divides it by two. The transfer function for the signal conditioner block (which includes the sensor) is:

$$H_{total} = H_{sensor} * H_{voltage\ conversion}$$
$$= 1\ volt/inch * 0.5\ volts/volt$$
$$= 0.5\ volts/inch$$

Therefore, at the 0.9 inch mark after 0.1 seconds, 9.1 inches from the required position, the sensor's output changes from 0 volts to 0.9 volts and the feedback signal is half that (0.45 volts). At the error amplifier, the setpoint hasn't changed, so the newly calculated error is:

$$e_e = e_{in} - e_f$$
$$= 5\ volts - 0.45\ volts$$
$$= 4.55\ volts$$

The new velocity with which the bed will move to correct the error depends on G, which we found to be 1.8 inches per second/volt, so the new velocity will be only:

$$V = G \times e_e$$
$$= 1.8\ inches\ per\ sec./volt \times 4.55\ volts$$
$$= 8.19\ inches\ per\ second$$

Which will still require slightly more than a second to close the gap! Table 6.1 shows that if the example servosystem corrected its error signal every 0.1 seconds, then the output speed would decrease each time. Even after 3 seconds, the servosystem still requires slightly more than a second to close the gap at its speed at that time!

The *actual* response of even the simple servosystem we have just examined would be quite a bit different from the description. In our example we ignored the effects of **friction** and **inertia.** Inertia would prevent the instantaneous acceleration that we assumed, and would tend to cause the machine bed to carry on past the desired position. Friction would reduce the total torque available to move the load to the goal but would help to prevent the bed from overshooting its desired position. The reader also may have noted that the motor in our NC system was theoretically capable of 600 rps (36,000 rpm)! Obviously we also ignored the upper limits, or **saturation** values, for the components.

We will examine friction, inertia, saturation, and other factors that affect the response of servosystems, and what can be done to overcome these inherent difficulties, in the following sections.

It is important to remember that *control can't simply be imposed on a system*! Figure 6.26, later in this chapter, shows that the characteristics of the controlled system

TABLE 6.1   RESPONSE OF THE EXAMPLE CONTROL SYSTEM IN FIGURE 6.3

Setpoint Position	Time	e_{in}	e_f	e_e	Speed	Position After 0.1 Sec.
0	-0.1	0	0.00	0.00	0.00	0.00
10	0.0	5	0.00	5.00	9.00	0.90
10	0.1	5	0.45	4.55	8.19	1.72
10	0.2	5	0.86	4.14	7.45	2.46
10	0.3	5	1.23	3.77	6.78	3.14
10	0.4	5	1.57	3.43	6.17	3.76
10	0.5	5	1.88	3.12	5.62	4.32
10	0.6	5	2.16	2.84	5.11	4.83
10	0.7	5	2.42	2.58	4.65	5.30
10	0.8	5	2.65	2.35	4.23	5.72
10	0.9	5	2.86	2.14	3.89	6.11
10	1.0	5	3.05	1.95	3.50	6.46
10	1.1	5	3.23	1.77	3.19	6.78
10	1.2	5	3.39	1.61	2.90	7.07
10	1.3	5	3.53	1.47	2.64	7.33
10	1.4	5	3.36	1.34	2.40	7.57
10	1.5	5	3.78	1.22	2.19	7.79
10	1.6	5	3.89	1.11	1.99	7.99
10	1.7	5	3.99	1.01	1.81	8.17
10	1.8	5	4.08	0.92	1.65	8.33
10	1.9	5	4.17	0.83	1.50	8.48
10	2.0	5	4.24	0.76	1.36	8.62
10	2.1	5	4.31	0.69	1.24	8.74
10	2.2	5	4.37	0.63	1.13	8.86
10	2.3	5	4.43	0.57	1.03	8.96
10	2.4	5	4.48	0.52	0.94	9.05
10	2.5	5	4.53	0.47	0.85	9.14
10	2.6	5	4.57	0.43	0.78	9.22
10	2.7	5	4.61	0.39	0.71	9.29
10	2.8	5	4.64	0.36	0.64	9.35
10	2.9	5	4.68	0.32	0.58	9.41
10	3.0	5	4.70	0.30	0.53	9.46
(inches)	(volts)	(V)	(V)	(V)	(in.sec)	(in.)

affect the system's response just as much as the characteristics of the controller do. A properly-designed control system interacts with the physical system being controlled, so that the combined system characteristics will be under control.

6.2 SERVOSYSTEM ANALYSIS

There are two ways of describing and analyzing the behavior of servosystems. We may describe the servosystem by its **transfer function** or by its **observed response characteristics**.

6.2.1 Analysis by Transfer Function

A transfer function (T.F.) shows the relationship between a system's (or component's) input signal and its response (output).

$$\text{T.F.} = e_{out}/e_{in}$$

The transfer function for an **open loop** control system is relatively simple. Each of the individual component transfer functions are combined to yield a single value for G, so:

$$e_{out} = G * e_{in}$$

and the open loop transfer function is simply:

$$\text{T.F.} = e_{out}/e_{in} = G$$

In a **closed loop** control system, however, the transfer function is more complicated because the output of the system is dependent not only on the input and the gain, but also on the measured output. The overall system transfer function of a servosystem like that in figure 6.2 includes the effect of the feedback loop function, H, and is:

$$\text{T.F.} = e_{out}/e_{in} = \frac{G}{1 + GH}$$

Techniques used to describe the behavior of a closed loop control system sometimes require that the **loop be opened** by disconnecting the feedback signal from the error amplifier. A transfer function that describes the **feedback signal's** opened loop relation to the input signal would include the signal conditioning:

$$\text{T.F.} = GH$$

where

$$\text{T.F.} = e_f/e_{in}$$

The transfer function approach, where the servosystem is described by a **mathematical function** (or **"model"**), is the approach we took in examining the servosystem in figure 6.3. There are several reasons why this type of approach is often avoided. In our example, we neglected so many factors that the formula developed was almost useless! We neglected the effects of friction and inertia in the system, we assumed that the speed of the motor was linearly related to the voltage, we ignored component saturation values, and we failed to acknowledge that the characteristics of the system may change with time! Figure 6.4 demonstrates some of the factors that we neglected in the gain (G) portion of our model, and some other factors that might be needed to describe better control systems.

A good transfer function of the NC control problem would require a very complex formula. The formula would include **"second order"** terms. Second order terms are terms such as **integral and derivative functions,** and a formula containing such terms is difficult to manipulate. It is standard to convert such formulas to more manageable forms through the application of a **Laplace transformation.** The Laplace transformation takes a time-based function and converts it to a frequency-based function, which can then be algebraically manipulated more easily. The rearranged formula can be converted back into a time-domain formula by the reverse Laplace transformation, but is

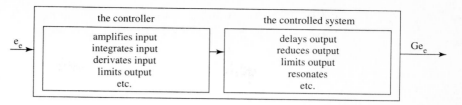

Fig. 6.4 A look into the G box of the transfer function

often left as a Laplace transform to evaluate the frequency response characteristics of the system. The term *transfer function* is often taken as implying that the formula is in Laplace transformed form. We will not discuss the Laplace transformation process in this book.

For most servosystem designs, only a rudimentary transfer function is worked out. The resultant approximation simplifies the system design, then the system is built and modified at a lower cost than developing an accurate transfer function.

There are situations where it is worth the time and money required to develop a good mathematical model of a servosystem.

It may be desirable to develop a mathematical model after the system has been built. For example, the characteristics of an existing airliner can be measured and a mathematical model of the airliner's response can be programmed into a computer. The model can be used in pilot training, and may also be used to predict the behavior of the system under conditions that it would be impossible, expensive, or dangerous to subject the actual system to (such as during engine failure).

Predicting the behavior of a proposed system to **rare conditions** is another area where a transfer function can be invaluable. Just how would a space station respond to the impact of a pea-sized meteorite? In such cases it is necessary to collect as much information about those conditions as we can, use the information to create a transfer function model of the proposed systems, then computer test the model. Only after the testing is done can large investments in time and lives be risked on the project.

On a level more immediately relevant to those involved in automation control, transfer functions are used in robotics. They are used in robot controllers to **calculate optimum actuator outputs** to achieve desired motions. Work is being done to improve on these often inadequate transfer functions so that they will eventually be used to improve controlled path motions (in which the route of the object in the robot's gripper is controlled). Robots in the future will then be capable of calculating their own sequence of operations in an unexpected situation.

Transfer functions are also used in **adaptive controllers.** Some adaptive controllers predict the result of an actuator's action, then use *prediction as feedback* in the control system. In some of these adaptive controllers, when the sensory information describing the actual response of the system is received, it is used to change and improve the model! In this way, the model changes as the characteristics of the physical system change.

Transfer functions must describe the response of a system to input signals and to environmental disturbances, so they must include accurate descriptions of the controller and of the controlled system. We will discuss servosystem response in section 6.3.

6.2.2 Analysis by Graphical Methods

Servosystems have been around longer than digital computers have been in common use, so it should be no surprise that non-numerical methods exist to describe the functions of servosystems. Graphical depictions of the response of a system (or of its components) enable designers and operators to understand the system, so that appropriate control techniques or settings may be chosen. Many servosystem components, particularly of the hydraulic type, include graphics depicting their response characteristics in their technical literature.

We will use the graphical approach to explain the interrelation among the three variables in servosystem design. The variables are: the **input signal** (or **disturbance**), the **physical system,** and the **control system.** Remember that no one or two of these variables is sufficient to explain the output of a servosystem. All three must be identified in the explanation.

Figures 6.5, 6.6, and 6.7 show three methods of graphically depicting the response of a servosystem to a changing input signal.

In figure 6.5, *position of a load is plotted against time* after the command system has suddenly changed the position setpoint (dotted line). The motor cannot change load position instantaneously, but must first overcome inertia. The inertia in the motor and load may cause the load to slightly **overshoot** the setpoint. Friction will absorb energy, so the output will *settle*, perhaps after some oscillations. Because of friction, the load may stop with a **steady state error** difference between the setpoint and the final position.

Figure 6.6 (a) uses **Bode diagrams** to show the *response of a system to various frequencies of repeating input signals or repeating disturbances*. Bode diagrams may be used to show closed loop response of servosystems, or may be used to show open loop response of a component or a system. (It is sometimes useful to disconnect the feedback signal of a servosystem, then determine the Bode diagram for the opened control system.)

The horizontal scales of a Bode diagram are logarithmic frequency scales. The upper plot of the Bode diagram shows the gain of the system. Remember that gain is the ratio of output to input. A gain of 1 means that the output of the system is exactly as the setpoint signal requests (or, one times the correct output value). A gain of less than one means, of course, that the response is less than what it should be, and a gain of more than one means that the system is overresponding. Gain is also plotted logarithmically.

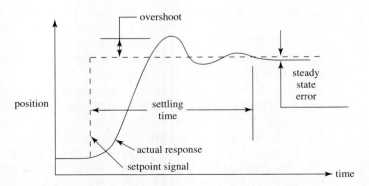

Fig. 6.5 Time-position plot of a servosystem output

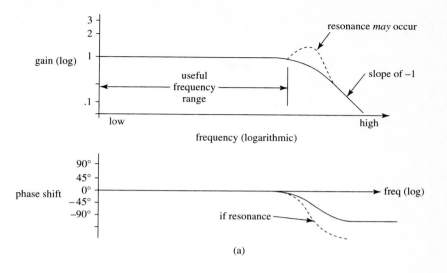

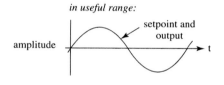

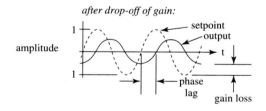

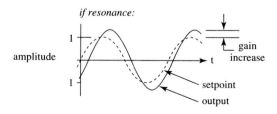

Fig. 6.6 Bode diagram. (a) The Bode diagram, showing gain and phase. (b) Time-amplitude plots of system input signal and output response, at three different input frequencies.

Other values for gain are sometimes used. Sometimes the **actual system gain** values are used (e.g., if an amplifier with a gain of 20 is used, "20" is used in place of "1" for full gain). Other times, **decibel (dB)** values are used. Zero dB gain means that the output is responding to the input as it should, while negative dB values mean underresponse, and positive values indicate overresponse. Decibels are ideally suited for logarithmic scales because they are in fact logarithmic values. (The decibel is not only a measurement of sound level.)

The second Bode diagram plot from the top shows *phase versus frequency values*. When the output of the system is keeping up with the input signal, it is "in phase." There is no phase lag or phase lead because the output value changes at the same time as the input value. At higher frequencies of the input signal, when the system can no longer change its output as quickly as required, the phase of the output *lags* the phase of the input signal.

The time-output response of *a simple servosystem* is often as shown in part (b) of figure 6.6.

At low frequencies, there is no error because it is easy for the servosystem to change the output as the setpoint changes slowly. The servosystem is useful at these lower frequencies.

At higher frequencies, the response time of the physical system (due to friction and inertia) causes a noticeable **lag in actual output** as compared against the setpoint. Since the peaks in the output are not reached until the setpoint value has reduced, and since a servosystem tries only to correct the difference between setpoint and actual output, then there is usually an **amplitude loss** in the output signal. This shows up as a loss of gain on the Bode diagram gain plot. With greater phase lag, the gain decreases more. At 90 degrees of phase lag, the peaks in the output occur while the setpoint value crosses zero, and there is no measurable output of the system in response to the rapidly changing input signal.

Every system and every system component has a **natural frequency** at which it resonates when disturbed. Even small amplitude, repeated input signals can result in gains of more than 1 at these frequencies. Resonance will be discussed later in this chapter when we discuss system stability.

If an *opened loop* Bode plot of a servocontrol system shows a gain of greater than 1 at a frequency where the phase lag is 180 degrees, the system will be unstable at that frequency.

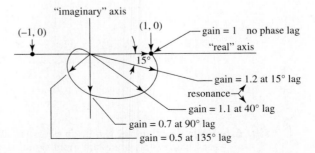

Fig. 6.7 Nyquist plot

Some servosystem component suppliers include Bode diagrams in their technical literature. Users of these components can ensure that the equipment they are buying responds to the frequencies anticipated in the application. Where it is desirable to know the response of a group of such components that will be assembled such that they are in series in the control system, Bode diagram gain values can be combined by multiplying the gain values and by adding the phase values at each frequency.

In figure 6.7, a typical **Nyquist plot** is shown. This is another graphical method of describing *the response of a component or of a servosystem to various input frequencies*.

The Nyquist plot is a curve drawn around an intersecting pair of axes. The curve is drawn by determining the response of the system at selected frequencies of input signal, then drawing vectors describing gain and phase shift for each frequency. All vectors are drawn from the same starting point. Each vector has a length proportional to the gain at the input frequency, and forms an angle with the horizontal that is equal to the output phase shift at that frequency. The Nyquist plot is generated by the endpoints of the vectors. (This is known as a time-domain Nyquist plot. Nyquist plots are also drawn using frequency-domain data obtained from Laplace transforms.)

The **Nyquist Stability Criterion** states that the closed loop system is unstable if the open loop Nyquist curve is outside the (−1,0) point, which would mean that the system has a gain of greater than one when the phase lag is 180 degrees.

Next, we will apply what we know about transfer functions and graphical methods of describing servosystem response to examining typical response characteristics of servosystems.

6.3 THE RESPONSE OF SERVOSYSTEMS

The response of the overall system depends on three factors:

- the type of input signal or disturbance
- the characteristics of the components of the servosystem
- the type of controller system selected

These three factors must be carefully controlled so that the system can be brought under control.

6.3.1 Stable Versus Unstable Response

If a system goes drastically out of control, it is said to be unstable. A stable response can also be an undesirable response! For example, if a system totally fails to respond to setpoint changes, its response is very stable.

Figure 6.8 (next page) demonstrates the difference between two types of stable and two types of unstable response.

A system that exhibits **monotonic stability** (fig. 6.8 (a)) causes its output to change toward the setpoint without overshooting that setpoint. This is also known as "first order" response. If the system is "critically damped," then the output moves toward the desired output quickly. If the system is "overdamped," then it takes too long to reach the desired output state.

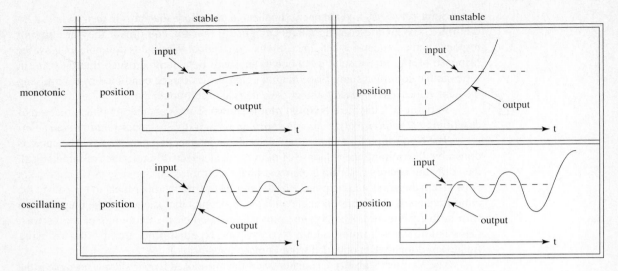

Fig. 6.8 Stable and unstable system response: (a) monotonic stability; (b) oscillating stability; (c) monotonic instability; (d) oscillating instability.

One source of overdamping might be high dynamic friction in a position control system. A more powerful actuator would overcome the friction, making system response quicker.

A system will exhibit **oscillating stability** (fig. 6.8 (b)) if it is "underdamped." Underdamped systems overshoot their setpoint. Overshoot may be due to inertia or to energy being stored in the system in another way. Systems that store energy are often called "second order" systems. If the magnitude of the overshoots does not increase, the system is considered stable. In a well-tuned controlled system, each overshoot should be less than the previous one.

Systems are often designed to have a small overshoot in the interest of fast response. Oscillating overshoots can be reduced with dynamic friction, or by dissipating stored energy in some other way.

Monotonic instability (figure 6.8 (c)) can occur as a result of poor system design or of system failure. Output will "run away" as shown if the feedback signal has the wrong polarity, or if the actuator drives in the wrong direction when trying to correct detected error.

If a short circuit causes a motor to run at full speed, or if a failed sensor results in the controller permanently trying to correct a problem that does not exist, then a system will go unstable.

Oscillating instability (figure 6.8 (d)) is a little more complex than monotonic instability. The system does not simply run away in one direction. It tries to correct errors by driving too hard, overshoots the setpoint, then corrects even harder in the other direction. Each overshoot is greater than the one before. The oscillations typically occur at the resonant frequency of the system or of one of its components.

One solution is to increase system damping to attempt to dissipate the energy as the system races towards the setpoint. Other solutions will be discussed in the following sections.

6.3.2 Input Signals, Disturbance Types, and System Response

A closed loop control system responds to error. Error is calculated by subtracting feedback from the setpoint. Error, therefore, changes if the setpoint changes. Error also changes if an external disturbance causes changes in the feedback signal.

The input signal, or **setpoint,** used to tell a servosystem what is expected of it can come from a variety of different sources.

- The input signal may be provided by **an operator** who adjusts a dial at a control panel.
- In some early automation applications, the manual operations of highly skilled operators were taped on **magnetic tape** (as they, for example, adjusted dials to produce a product). These tapes were played back and used as setpoints for servosystems that replaced human operators.
- **Analog sensors and analog computers** have been used to provide setpoints for servosystems.
- **Digital computer** control is now common. A computer is programmed to provide sequential digital numbers that are used as setpoints. A digital setpoint in a Computer Integrated Manufacturing system may originate at, for example, a CAD drawing, and may be used to position a cutting tool in an NC controlled lathe.

A **disturbance** changes the error by changing the sensed output of a controlled system. The disturbance can come from one or many sources. Environmental conditions are often sources of disturbances. For example, room temperature can change as outdoor temperature changes. Increased weight on a conveyor belt can slow the belt. Electrical "noise" may change the feedback signal, even if the system output hasn't actually changed.

Disturbances may have their sources *within the control system*. The **characteristics** of controlled systems may **change over time.** Lubricants, for example, can warm up, reducing friction. Electronic component characteristics can drift over time.

A common type of error signal change is where the error signal changes level just once. This is known as a **unit signal** change. Four variations of unit error signals are demonstrated in figure 6.9.

The simplest of these unit signals is the **step** signal. The step signal is just that: the setpoint or feedback suddenly changes from one value to another, resulting in a new error signal level. An example of a step signal was used in the NC position control example (figure 6.3). The setpoint used to move the table was suddenly changed to demand that the table move from the 0" position to the 10" position.

Figure 6.10 shows a possible response of a position control servosystem to a unit step input signal. Note that the physical system cannot respond with instantaneous acceleration, nor with infinite velocity to the suddenly changed setpoint. There is a lag before the position even begins to change, as the mass is accelerated. There is a limit on the velocity reached. There may or may not be an overshoot. The terminology shown on this figure is in common usage in control theory, and we will see it again.

Resonance in response to a unit step error signal may occur under certain circumstances. If the energy storing components (e.g., inertia, capacitance) in the system store energy faster than energy dissipating components dissipate energy (e.g., friction,

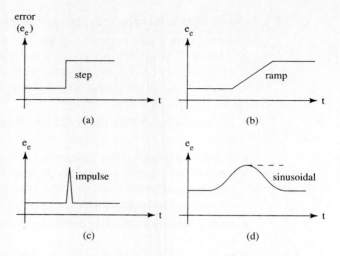

Fig. 6.9 Unit error signals: (a) step; (b) ramp; (c) impulse; (d) sinusoidal.

resistance), an error signal change can result in repeating oscillations. If the gain is high enough, the system may be unstable.

If the feedback signal lags the actual changes in the output, then the calculated error will be incorrect. If the feedback is delayed long enough, the actuator may even drive the output further past the setpoint than the original error was, so oscillating instability will occur. Delayed feedback may be due to delay between when the actuator causes a change, and when that change reaches the sensor (e.g., a controlled valve allows material onto a conveyor that carries it to a bin, but the sensor in the bin can't detect the future bin contents until much later, when the material actually reaches the bin). Delayed feedback may also be due to the distance the feedback signal must travel (e.g., radio communications for controlling spacecraft).

Another unit signal is the **ramp.** A ramp input signal might be provided in place of a step signal if the step signal might cause oscillating instability. With a ramp positional setpoint signal, output may be able to keep up with setpoint changes so that velocity as well as position can be controlled.

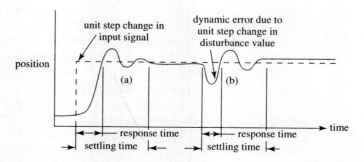

Fig. 6.10 Position control response to: (a) a unit step input signal; (b) a unit step disturbance

A unit **sinusoidal** input signal, in which the slope of the signal changes, might be used for control of acceleration in a position control application. Assume that position control is used on a milling machine, where the milling cutter is stationary and the workpiece feeds past it. The position controller for the machine bed provides a sinusoidal signal to cause the workbed to gradually accelerate to the right, for a single pass past the cutter, gradually decelerate, accelerate to the left for another cutter pass, and then decelerate to a stop.

The **impulse** type of unit signal is usually an undesirable disturbance signal. The source of this type of signal might be a large motor, or an arc welder, with a large changing current draw near the circuitry through which the servosystem setpoint or feedback signal passes. The electromagnetic interference from the motor or welder would produce a large voltage spike in the signal. Digital to analog converters are also sometimes responsible for impulse signal generation. If a servosystem positions a workpiece to be drilled, any response by the position control system to an inadvertent impulse input signal might snap the drill bit off in the workpiece. Impulse signals are often of short duration, and many systems can't respond fast enough to such signals to do any damage. If impulse disturbances are a problem, a filtering circuit must be included in the control system to prevent the system from responding.

Figure 6.11 demonstrates the various types of **repeating** signals that a servosystem might be expected to respond to. The response of a controlled system to a repeating input or disturbance signal can vary with the frequency of the repeating signal.

Steady state error (see figure 6.5) can cause the output of a system to lag a repeating input. The result is called **"following error"** or **"phase lag."** As discussed in the section on Bode diagrams, when there is a phase lag, the peaks in the output occur after the peaks in the input signal. As the output approaches its peak, the error signal will already be trying to drive it in the reverse direction, so the peaks in the output may not reach the magnitudes they should. This reduction in amplitude of output is called **"gain loss."**

Systems responding to repeating input or disturbance signals also can exhibit **oscillating instability.** The instability may be due to the reasons described earlier in the

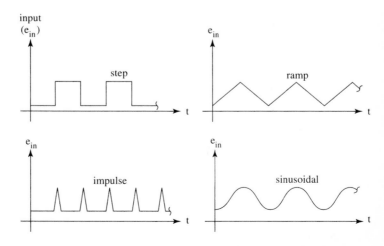

Fig. 6.11 Repeating input signals

discussion on oscillating instability (in section 6.3), or may occur under the following conditions.

If the signal frequency is at the natural frequency (resonant frequency) of the system or of one of its components, and if there is insufficient damping in the system, each output oscillation will have greater magnitude than the previous oscillation. Oscillating instability will also result if the feedback signal lags the setpoint signal by approximately 180 degrees. The feedback signal sign will therefore be opposite to the input sign, and after subtracting feedback from setpoint, the error signal will then be greater than the setpoint! The system will respond, using more force than required.

There are several reasons why feedback could be delayed. There may be a delay between when the output is measured and when a new output is calculated and delivered (e.g., a PLC reads a sensor's output, but takes 15 milliseconds to calculate and output a new actuator control value). The delay may be between the time that an actuator causes a change in output and when the change reaches a sensor (e.g., grain is released onto a conveyor belt by a controlled valve when a sensor detects that grain level is low in a grain hopper; despite immediate valve response, the grain requires 10 seconds of travel time on the conveyor before it reaches the grain hopper and the sensor). Or, the sensor may be poorly placed. It may be sensing a condition that is not the true output. (For example, an elastic drive belt is between a motor and a sensor. The motor makes a correct position change at its end of the belt, but the change doesn't reach the sensor until the elastic belt finally transmits the energy. In fact, at the belt's resonant frequency, the sensor's end of the belt might be moving in one direction while the motor is driving the other end of the belt in the other direction.)

Oscillating instability problems can be solved by one or more of the following approaches:

- Add damping to dissipate stored energy.
- Make the controlled system "stiffer," effectively increasing its natural resonant frequency (e.g., replace an elastic component with a component that deforms less in response to the same forces).
- As a general rule, no servosystem should be expected to respond to a signal that is changing at a frequency more than half the resonant frequency of any of the components. This will ensure that no one component is responsible for causing the failure of the whole servosystem. Component resonant frequencies must be obtained from spec sheets, or by vibration testing.
- The sensor might be moved closer to the actuator's output than to the system's output.
- The source of delay might be reduced so that input to feedback phase lag is less than 90 degrees. As a poor alternative, delay may be *increased* so that feedback lags input by about 360 degrees. With sufficient damping, this method can eliminate oscillating instability, even if it increases following error.

6.3.3 Physical Characteristics and System Response

We have already discussed some physical characteristics of controlled systems (elasticity, sensor placement, etc.) in the previous section. This section will discuss other characteristics one might observe in controlled systems. You may wish to refer to figure 6.27, which shows a complete servocontrolled system including the physical characteristics discussed in this section.

We know that a real system has **friction** and **inertia.** In the NC position control system depicted in figure 6.3, friction and inertia would prevent the controller from instantaneously changing the position of the worktable. In the design of control systems for industrial robots with DC motors, it is recognized that motor **torque** (not speed) is controlled by controlling current. This is shown in the following formula:

$$T_m = K_m i$$

where:

T_m = motor torque
K_m = a constant for the motor
i = DC current supplied to the motor (controlled by the servosystem)

This theoretical torque output is reduced by the reactionary torque due to friction, and only the reduced torque is available to cause acceleration. The actual acceleration is further reduced by the inertia of the system. Figure 6.26 and the following equation show how actual acceleration is dependent on friction and inertia:

$$a = \frac{T_m - F}{I_m}$$

where:

a = the angular acceleration of the load
F = the opposing torque due to (dynamic) friction in the system
I_m = the mass moment of inertia of the system

Figure 6.26 also demonstrates that the final output position is the result of integrated (accumulated) velocity, which is itself the integral of acceleration.

What is not immediately obvious in the equation, or in figure 6.26, is that *friction and inertia may change with time*!

In a robot, inertia in a swinging arm is different when the arm is fully extended than when it is held in tight to the axis of rotation and, of course, varies as the robot picks up and releases heavy workpieces. Friction varies as the lubricants heat up or get used up. When the arm accelerates from a stationary position, **static friction** ("stiction"), which is higher than dynamic friction, must be overcome before the arm starts to move.

The effects of friction and inertia are often adjusted intentionally, or the gain of the servosystem may be modified, so that **damping** is within reasonable limits. Increasing friction increases damping. Adjusting the amplifier gains in the controller effectively alters system damping.

Other system characteristics may affect system response. Many control systems (and components in control systems) have a **deadband**, i.e., a range of small input values to which they do not respond at all. This deadband may be intentionally built in, as in a hydraulic valve in which the spool must be moved a measurable amount from the center before it starts to open a port, or it may be an unavoidable characteristic, such as in a motorized system where the motor doesn't respond to a small DC voltage because the torque generated is less than stiction.

Figure 6.12 shows the fluid flow versus current curve for a proportional servovalve. Flow should be precisely proportional to the electrical current, but note that deadband due to the spool and port construction prevents any fluid from flowing at low current levels.

Somewhat similar to deadband is the **hysteresis** effect found in most components. A measure of hysteresis would be the measure of energy lost in reversing some prior action.

A light switch has hysteresis intentionally built in. When the light is turned on by a human who lifts the switch lever, the light switch lever does not fall back when the lifting finger is removed. The expenditure of new energy is required to return it to its previous state. Figure 6.12 shows the effects of hysteresis in a hydraulic servovalve. Note that as current to the valve increases from zero, the amount of fluid flow is read from the upper lines. If the current stops increasing, then decreases, the resultant fluid flow holds steady for a while despite the reducing current, then reduces as read using the lower lines.

Obviously, the ability of a control system to accurately attain a given output is reduced if the sensor, actuator, and other components all have hysteresis. A small amount of hysteresis is considered a desirable characteristic in servosystem sensors, because it tends to counteract steady state error, and because it reduces the tendency of the system to oscillate near the setpoint.

There are limits on a control system or an individual component's response. **Saturation,** a term that is usually used with electronic components, means that the output is at its upper or lower limit.

An amplifier is said to be saturated if it is outputting its maximum value. At saturation, the quoted gain for the amplifier is irrelevant because any further increase in the input value cannot result in an increase in the output. A hydraulic valve can be "saturated" in the same way, when it is fully open. A motor turning a conveyor fitted with a speed governor can also be said to be saturated when the motor is driving the conveyor as fast as it will be allowed to go.

These limiting values, whether electrical, mechanical, hydraulic, or otherwise, reduce the responsiveness of a servosystem.

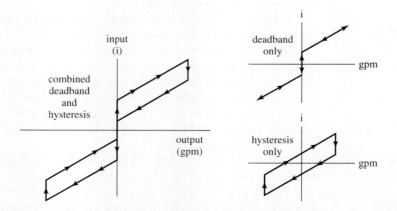

Fig. 6.12 Deadband and hysteresis in an electrically actuated hydraulic valve

The combined effects of the above-listed characteristics are often bunched together into system characteristics called **deadtime, capacitance,** and **resistance.**

Deadtime (not deadband!), shown in figure 6.13, is the lag between when a unit step input is received, and when the output of the system starts to respond. It is commonly due to actual distance between the location of the controlled variable and the point at which the servosystem acts.

In the grain hopper example used previously, the delay between when a valve allows grain onto a conveyor and when the grain arrives at a hopper where grain height is being controlled is deadtime. Deadtime is evident in the actuation time of other components as well, such as in the time it takes for a temperature sensor to warm up. Another source of deadtime is where a process is being monitored by a digital computer. The digital computer may be programmed to sample the output from the sensor at periodic intervals. When the sensor's output changes, the servosystem will not respond until the digital computer gets around to sampling it.

Capacity and resistance to change are interrelated in terms of defining the rate at which a servosystem can respond to a change in setpoint. The result is as shown in figure 6.13, where the term *capacitance* has come to be used for a *time constant, based on actual capacitance and resistance*.

Just as in electrical theory, the "capacitance time constant" means the amount of time it takes for the output of the system to reach 63.2% of the required output level.

Systems with high **capacity** require larger amounts of change to achieve a small measurable change. The larger the capacity, the slower the measured output will change. In the grain hopper example, the capacitance of the system goes up if a larger diameter hopper is used. It will take longer to change the height of grain in a large diameter hopper than in a small diameter hopper.

Resistance is found where something proportionally reduces the rate at which a servosystem can change its output. In the grain hopper example, if 10% of the grain on the conveyor belt consistently falls off as it is being conveyed to the storage bin, then this acts as resistance, reducing the bin filling rate.

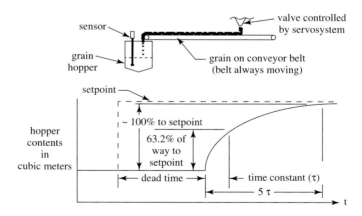

Fig. 6.13 Deadtime and capacitance time constant in the response of a servosystem to a unit step input

In a system with several sources of deadtime and capacitance, the individual deadtimes are additive. Capacitance time constants, when combined, yield a deadtime followed by a single capacitance time constant that is longer than any of the individual time constants.

6.3.4 Controller Algorithms and System Response

So far, most of our discussion of servosystems has assumed one simple type of controller: the type shown in figure 6.2. This controller's output is proportional to the error. There are five different types of algorithms used in controllers. They are:

- the **on/off** control type (including differential gap)
- the **proportional (PE)** type (the one we have been using)
- the **proportional-integral (PI)** type (can be used as just integral type)
- the **proportional-derivative (PD)** type
- the **proportional-integral-derivative (PID)** type

On/Off Control

In **on/off control,** the error amplifier has only *two possible outputs*, and is usually called a **comparator.** If the setpoint value is higher than the feedback signal, the comparator output is at maximum. If the setpoint value is lower than the feedback signal, then the comparator output is at its minimum (or most negative). An on/off controller example is shown in figure 6.14. It is similar to the type of heat control that exists in most homes, where the furnace turns on if the thermostat detects that the actual room temperature is lower than the set temperature, and turns off otherwise.

On/off control is *similar* to home temperature control because home temperature control is not simple on/off control, it is actually **differential gap** control. Typically, a thermostat set for 20 degrees might not turn the furnace on until the room temperature drops 1 degree *below* 20 degrees, and might not turn the furnace off until the room temperature rises to 1 degree *over* 20 degrees. The on and off points are therefore separated by a differential gap of 2 degrees. (This is also an example of intentionally applied hysteresis.) Differential gap control is far more common than simple on/off control, because it prevents the actuator from repeatedly turning on and off when the output is near the setpoint. Because the output can drift within the differential gap, differential gap control is sometimes called **floating point control**.

Proportional Control (PE)

We have spent much of this chapter discussing the characteristics of proportionally controlled systems.

The block diagram and an op amp–based circuit for proportional control are shown in figure 6.15.[1] A proportional amplifier with adjustable gain (K_P) is shown between the error amplifier and the sensor. Please remember that the sensor is more often placed after

---

[1] Operational amplifiers are covered in more detail in the next chapter. Readers familiar with op amps will recognize that the op amp–based circuits presented in this chapter can be built using fewer op amps. The author has chosen to present circuits that reflect the block diagrams.

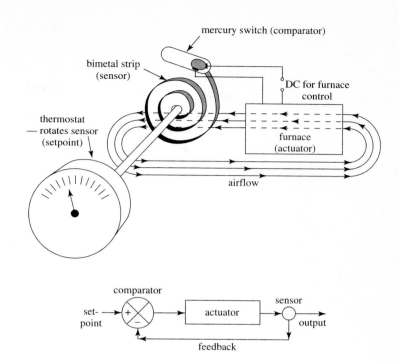

Fig. 6.14 On/off control block diagram and example

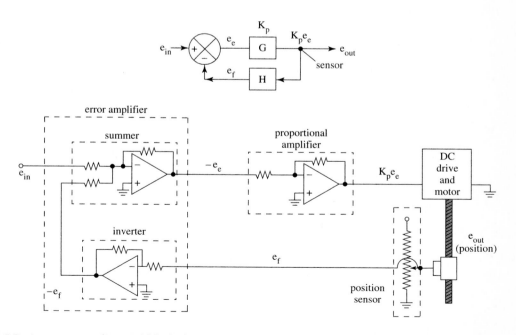

Fig. 6.15 A proportional control block diagram and op amp–based circuit

6.3 The Response of Servosystems **171**

an actuator in a real controlled system. The NC control example, previously discussed, is a more complete example of a proportionally controlled system.

The generalized equation for proportional control is:

$$e_{out} = K_P e_e$$

where:

K_P = the gain setting for the proportionality

In proportional control, the error signal from the error amplifier causes *a response that has a strength proportional to the error*. A position control system that is asked to move a long distance will respond harder than if asked to move a short distance.

With a PE control system, *the response begins as soon as the error signal becomes non-zero*. The measured output of a proportionally controlled system might look like figure 6.16 in response to a unit step input signal change.

The resultant move may exhibit some dynamic error, called **overshoot**, especially if the system is underdamped as it will be when moving a heavier than average load. If there is overshoot, the servosystem controller will respond by driving the output back toward the setpoint again, resulting in an overshoot in the opposite direction. These oscillations about the setpoint will eventually damp out, but until then the system will oscillate at a **resonant frequency** characteristic of the loaded system.

The output will not stop exactly at the desired condition. As it approaches that condition, the error becomes smaller. Eventually, the error signal becomes so small that, even after proportional amplification, the control system output is too small to overcome opposing forces (such as friction). Attempts to reduce this small final error, called **steady state error** or **residual error,** by increasing the gain will only cause more overshoot error.

Figure 6.15 also shows an op amp–based proportional controller, with three op amps: two in the error amplifier, and the third as the proportional amplifier. Remember that op amps amplify the difference between their two input voltages. In this circuit, all op amps have their non-inverting input grounded, so the difference between the two inputs *is* the voltage at the inverting input. The output of each op amp is its input value, amplified and inverted.

The op amp used as the *summer in the error amplifier* receives both the setpoint voltage and the (*inverted*) feedback voltage. The summer's input voltage is the difference between the two. The op amp–based error amplifier outputs a signal proportional to the difference between the setpoint and feedback without drawing significant current from either circuit.

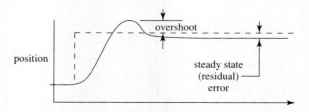

Fig. 6.16 Response of a proportionally controlled system

The op amp connected as a proportional *amplifier* is where the proportional gain comes from. The gain of an op amp without feedback is a high but unstable value. A stable gain factor can be chosen by selecting a feedback resistor (R_f) and input resistor (R_i) where:

$$\text{GAIN} = K_P = -(R_f/R_i)$$

The proportional gain op amp re-inverts the voltage value previously inverted in the error amplifier, so final voltage output is directly proportional to the difference between the setpoint and the feedback condition (if no op amp is saturated). In our circuits, this output voltage is sensed to provide feedback to the error amplifier, but normally it would be used to (for example) control a power supply for a DC motor, and the position of the load being moved by the motor would be sensed to provide the feedback.

In fact, few proportional controllers provide output proportional to error over the whole range of possible error values. For strong control, a proportional controller should respond strongly (high KP) to small input signal changes. Limits on the output of the controller prevent proportionality at large input signal changes. The concept of **Proportional Band (PB)** is used to include the limits of proportionality. A controller's proportional band is defined as:

$$PB = \frac{\text{The percentage of the span of input values that is proportional with output values}}{\text{The percentage of the span of output values that is proportional with input values}} \times 100\%$$

where:

$$\text{span of values} = \text{difference between the highest possible value and lowest possible value.}$$

Proportional Integral Control (PI)

A proportional integral controller, sometimes called a "reset" controller, is *intended to solve the problem of steady state error.* The generalized equation for proportional-integral control is:

$$e_{out} = K_P\, e_e + K_I \int e_e\, dt$$

where:

K_I = the gain setting for the integrator, and

$\int e_e\, dt$ = the sum of the error values to the current time

Figure 6.17 shows the block diagram for this type of control, and how such a control system can be built using op amps. Note that this type of control *includes proportional control*, so the controller can still respond to an error between the setpoint and the actual measured output of the system by driving (for example) a DC motor proportionally to the amount of error.

When this controller first receives a unit step input, the error signal immediately becomes non-zero. When amplified by the proportional gain, this immediately causes a motor torque to be generated. The integral of the error takes time to build up. While it is very small, even amplified by the integral gain, it does not contribute much to help drive the DC motor. After the proportional amplifier has caused the load to move very close to the position required by a setpoint, the error signal becomes small, so even when amplified by a proportional gain, it cannot contribute enough current to cause the DC motor to close the small remaining steady state error. By now, the integral of the error signal will have grown to a respectable value, and *will continue to grow as long as the steady state error exists*. This integrated value, though amplified by even a small integrator gain, can cause sufficient current to drive the DC motor to close the steady state error. Note that the integrated error value *does not go to zero when the error becomes zero*, so the motor will continue to output a torque. This torque may be required to hold a load against gravity. If the torque is high, *it may cause overshoot* and thus generate a negative error. Integrating negative error with the positive integrated error value will then decrease the integral value until only the required small holding torque is being generated.

The op amp circuit in figure 6.17 is identical with that in figure 6.15, except that some extra output current is coming from an op amp that has a capacitor in its feedback circuit. *This op amp integrates the input voltage it receives*, and outputs a value proportional to the sum. A fourth and fifth op amp are now required. The fourth op amp sums the

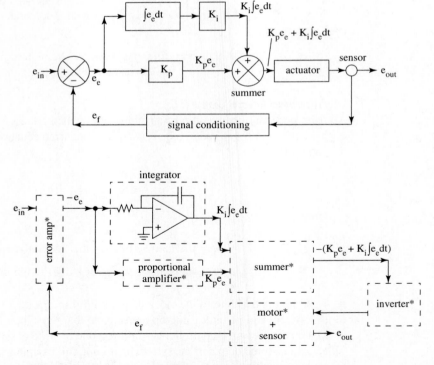

Note: Details for components marked with an * are shown in Figure 6.15.

Fig. 6.17 Proportional integral control block diagram and op amp–based circuit

voltages from the proportional op amp and the integrating op amp. Remember that each op amp inverts its output, so the fifth op amp is required only to invert the negative output of the fourth.

As a general rule, *the gain on the integral function should be kept small relative to the proportional gain*. If large error values are expected, it may be necessary to adjust the circuit to further reduce the rate at which the integrator output is allowed to grow.

In "process control" applications, a setpoint is rarely changed. Error values are usually due to gradual changes in environmental conditions. In such applications it may be acceptable to dispense with the proportional control component and depend entirely on the integral controller. This variation on a PI control system is called an **integral controller.** In process control jargon, integral control is often called **reset control,** because it resets the controlled output to its previous value after it has been disturbed.

The integrator constant K_I, which has been treated as a unitless multiplier so far in this discussion, is sometimes quoted in terms of **"repeats per second."** The output of an analog integrator increases by an amount equal to its input (it repeats) after one time constant. The time constant in seconds is found by multiplying the values of the resistor (ohms) by the capacitor value (farads). The integrator constant is also sometimes described as **"reset time (Tn)"** in minutes or seconds, which is actually the inverse of K_I.

$$K_I = 1/RC \text{ (repeats per second)}$$
$$Tn = 1/K_I = RC \text{ (seconds)}$$

Because the addition of the integral control makes response time shorter, *PI controlled systems typically have resonant frequencies about 50% higher than do PE controlled systems.*

Proportional Derivative Control (PD)

Proportional derivative control, sometimes called "rate control," *modifies the rate of response of the servosystem in order to prevent overshoot*. PD controllers are rarely found in industry, although derivative control is part of PID control (to be discussed in the next section). We will discuss PD controllers because the effect of the derivative function is easier to understand if we do not have to include the effects of integral control as in PID control.

There are *two methods of applying the derivative control* portion of PD control; both methods are used *to prevent overshoot.*

One form of PD control uses the derivative term to increase the rate of response of a servosystem to a changing error signal. This type of PD control (shown in figure 6.18 and figure 6.19) is described by the following equation:

$$e_{out} = K_P \, e_e + K_D \frac{d(e_e)}{dt}$$

where:

$$K_D = \text{the gain on the derivative term, and}$$
$$(d(e_e)/dt) = \text{rate of change of error}$$

In this PD control system, an increasing error signal *increases* the system output. This type of PD control system is sometimes called **predictive** control because it causes the system to work harder to correct an error if the error is growing at a fast rate.

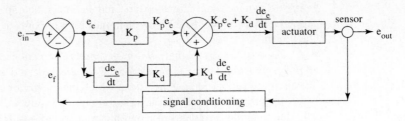

Fig. 6.18 Block diagram for a proportional derivative controller

The presence of the derivative term means that the proportional gain can be kept lower because the derivative value will cause acceleration as error grows. The tendency of the system to overshoot is further reduced because, as the error reduces, a negative derivative results! Adding this negative value to the still positive proportional value will reduce the output before overshoot can occur.

An alternate form of PD control is described by the block diagram in figure 6.20, and in the following equation:

$$e_{out} = K_P \, e_e - K_D \frac{d(e_{out})}{dt}$$

where:

$$\frac{d(e_{out})}{dt} = \text{rate of change of the output}$$

Note that in this formula, the output of the system is *reduced by a factor depending on the rate of change of the output*. This servosystem might control the position of a machine bed. As the machine bed picks up speed on its way to a new position (i.e., the rate of change of position increases), the derivative signal increases. Subtracted from the output in the summer, the derivative signal *reduces* the system output, slowing the system down as it reaches its goal. The proportional component will still attempt to correct error in the position of the machine bed, but the derivator will act as a limiter on the *velocity* of the machine bed. The net result is that, since the machine bed builds up less momentum, the tendency to overshoot will be reduced.

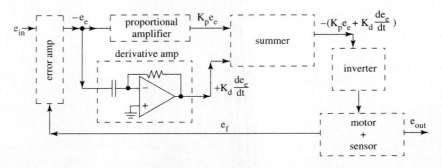

Fig. 6.19 A proportional derivative control circuit using op amps

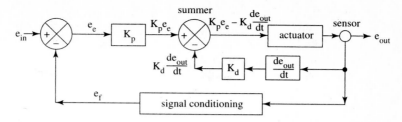

Fig. 6.20 An alternate proportional derivative control system block diagram

Figure 6.21 shows how this PD control system can be built using op amps. This circuit includes the op amp circuit for PE control. The capacitor in the derivative op amp circuit will allow a current to the op amp input only while the sensor's voltage is changing, so the derivative op amp's output will be proportional to the rate of change of the system's output. As in the PI circuit, the presence of the proportional op amp ensures that there will be some system response whenever there is an error signal.

In either type of PD control system the gain on the derivative function (K_D) is usually kept low. This is because the derivative term tends to inhibit the response of the system as error reduces. It can even result in a reversal of controller output before the desired output is achieved!

There are some conditions where servosystems with derivative control do not respond well. In *noisy environments* (electrical noise can occur if wiring is inadequately shielded), there will be rapidly changing voltage signals in the inputs to the derivative op amps. The derivative terms will respond by going very large as the spike rises, and again as the spike falls, causing the controller to respond by driving back and forward very rapidly. Thus, derivative functions are usually avoided if it is known that large electromagnetic field sources will be nearby.

"Derivative time (Tv)," in minutes or seconds, is sometimes used instead of K_D to describe the amount of derivative control. Because the derivative term responds to increasing error by increasing the controller's output, the controlled system responds earlier to a unit ramp input signal. The time saved is the derivative time.

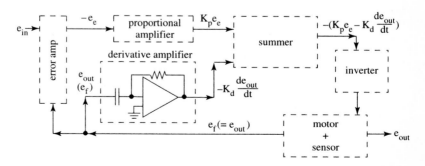

Fig. 6.21 An alternative proportional derivative control system circuit using op amps

Proportional Integral Derivative Control (PID)

Now that we have discussed PE, PI, and PD control systems, there is little new that can be said about Proportional-Integral-Derivative control systems. A typical PID controller will have some method of adjusting the gains on the proportional, integral and derivative terms, so that a balance between straight **proportional response, steady state reset ability,** and **rate of response control** can be found for the system being controlled.

There are two implementations of PID control, again based on how the derivative control is used. The most common block diagram for PID control is shown in figure 6.22, and a circuit for this type of controller is shown in figure 6.24. Note that the circuit is simply a combination of the circuits for PI and PD control. This PID controller, which can be set to optimize rapid response, without overshoot, and with steady state error correction, is described by the following generalized equation:

$$e_{out} = K_P \, e_e + K_I \int e_e \, dt - K_D \frac{d(e_e)}{dt}$$

The alternate method of connecting the derivative function can be selected. This alternate PID control type is demonstrated in block diagram form in figure 6.23, as a controller circuit using op amps in figure 6.25, p. 180, and in the following generalized transfer function:

$$e_{out} = K_P \, e_e + K_I \int e_e \, dt - K_D \frac{d(e_{out})}{dt}$$

Yet other forms of the above PID equations exist, and are quite commonly used. The equation:

$$e_{out} = K_M \left[K_P \, e_e + K_I \int e_e \, dt + K_D \frac{d(e_e)}{dt} \right]$$

includes a multiplier, K_M, that can be used to increase or decrease the controller output without altering the proportional-integral-derivative ratio.

A nearly identical PID equation, as used in Europe and written using DIN standard notation, looks like this:

$$y(t) = K_P \left[(X - W) + \frac{1}{Tn} \int (X - W) \, dt + Tv \frac{d(X - W)}{dt} \right]$$

where:

$$y(t) = e_{out}$$
$$(X - W) = e_e = (e_{in} - e_f)$$

PID control, once rare, is *rapidly becoming the industry standard* as **digital controllers** take over the function of analog components. This is not to say that op amps

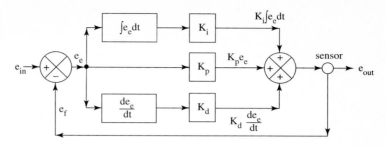

Fig. 6.22 Block diagram for a PID controlled system

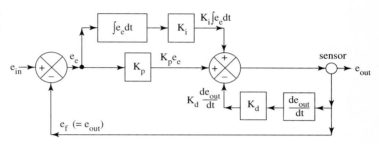

Fig. 6.23 Block diagram for an optional PID controller

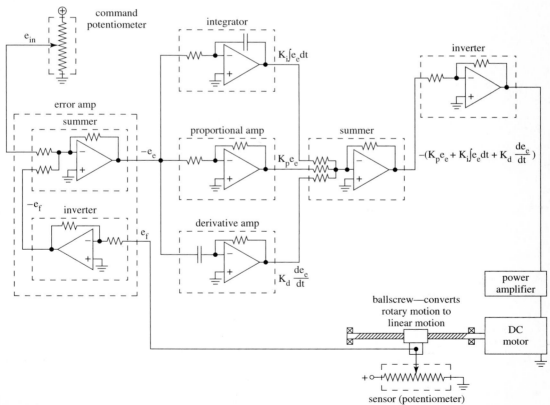

Fig. 6.24 PID control circuit using op amps

6.3 The Response of Servosystems

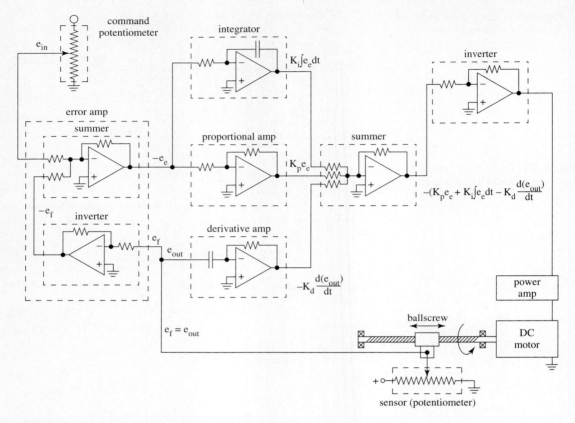

Fig. 6.25 Alternate PID control circuit with op amps

are obsolete! There is still the necessity to amplify the relatively tiny signals that digital computers output, and to sample DC levels in sensor circuits so the information can be "digitized" for computer usage. In fact, *analog control of subsystems that are under the supervisory control of a digital computer is still the norm*, because it allows the digital computer to spend its valuable time solving the difficult differential equations.

Tuning of PID Control

Purchased analog PID controllers are often supplied with user-accessible controls for altering K_P, K_I, and K_D. The following procedure can be used to tune an analog PID controller of the type shown in the block diagram in figure 6.22.

1) Adjust the integral and the derivative gain to zero.

2) Gradually increase the proportional gain from zero, applying a unit step input signal at regular intervals. In this manner, find the proportional gain value that will cause the system to respond with a constant-amplitude oscillation. Note the setting and the frequency of the oscillation.

3) Reduce the proportional gain to 1/1.7 of the value noted in step 2.

4) Increase the integral gain until a setpoint disturbance causes a system response with a frequency 50% higher than that observed in step 2.

5) Increase the derivative gain until the frequency of the system's response to a setpoint disturbance has returned to that observed in step 2. This should occur when derivative gain is 20% to 25% of the integral gain value.

6) If faster system response is required, and more overshoot is tolerable, increase the integral gain. This may require increasing the derivative gain as well.

7) If overshoot is excessive, decrease the integral gain, then decrease the derivative gain as necessary.

Digital PID controllers **can** be more difficult to tune. The section on digital controllers describes why this is so, but the tuning method described above shouldn't be expected to work with digital controllers. Digital controllers (microcomputer-based and PLC based) are most easily used where setpoint and feedback changes occur slowly. Digital controllers, on the other hand, can often be set up as constraint controllers (described in the next section), which can make tuning easier. Faster computers are reducing the differences between analog and digital controllers, to the extent that most servocontrollers are digital now. Most of the newer process controller modules even include **automatic tuning** features!

Developments in **artificial intelligence (AI)** have made a new generation of process controllers called **fuzzy logic** controllers available. To "tune" one of these controllers, the user must enter a description of the system being controlled. The fuzzy logic controller then calculates how best to control that system.

6.4 THE CONTROLLER AND THE CONTROLLED SYSTEM

In section 6.1, we noted that the response of the closed loop control system depended on three factors:

- the shape of the error signal
- the characteristics of the physical system
- the type of controller

Figure 6.26 is a model of a complete, closed loop, position control system. This model can be examined in order to review the material in this chapter so far, or it can be used as the basis for writing a computer program to evaluate a complete closed loop control system's response.

The model includes optional input values. An offset, or feedforward, value can be included at the error amplifier for adding to the error signal. A positive or negative disturbance value can be added at the same location, so that the error signal can be adjusted for environmental effects. The disturbance values may affect acceleration, velocity, or position, and their effects can be added where appropriate.

Any of the gains (except the drive amplifier, of course) can be set to zero, so the control system algorithm is selectable. The model recognizes that the derivative term may be based on rate of change of error, or of output, or of the setpoint, so a switch is shown to select one of those sources.

Values for Drive amplification, actuator characteristics, inertia, friction, and deadtime must be entered. (Remember that the inertia and friction values, and possibly even the deadtime value, may be time dependent.) The sensor and signal conditioner transfer functions, with a feedback gain adjustment, must be entered. Various types of input signals can then be "tried" to find the system's capacitance, resistance, total deadtime, and frequency response characteristics.

6.5 OTHER CONTROL CONFIGURATIONS

In addition to the control system algorithms (on/off, PE, PI, PD, PID) discussed so far, there are several alternate configurations of these five controller types, which we will now examine. The configurations examined will be:

- nested control loops
- direct synthesis control (DSC)
- adaptive direct synthesis control (ADSC)
- feedforward
- predictive control
- cascade control

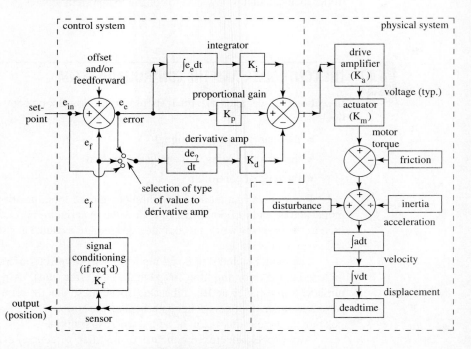

Fig. 6.26 A complete closed loop control model

- ratio control
- multiple output control
- constraint control

Nested control loops, as shown in figure 6.27, are commonly used where more than one characteristic of a system's output must be controlled at the same time. In this example, the primary function of the control system is to drive the load to a specified location, but there is an internal speed control loop to ensure that velocity of the system is under control during the move, and a third inner current-control loop that controls motor torque. This type of nested control is used in most robot controllers.

A good flexible manufacturing system (FMS) in a CIM environment would likely have several closed loops nested within other closed loops.

Direct synthesis control (DSC) is useful *where the time lag between a control system output and feedback reception is unmanageably long.* In such a case, it is often desirable to include special circuitry (or programming in the case of a digital controller) to *simulate the response of the system* and to provide timely feedback to the controller. The controller thus controls an actual system with its output, but receives feedback from the synthetic system. The DSC control technique is shown in figure 6.28.

DSC might be used for applications such as satellite guidance. A control system sends a command to fire a trajectory control jet. If the satellite is near Jupiter, there is a 2 hour message transmission delay, so the jet fires two hours after the command is sent. The controller can't get position control feedback from the satellite for another two hours. A computer can calculate the probable trajectory resulting from the firing of the jet within milliseconds of issuing the command. If the computer model's output is fed back to the satellite controller, we have DSC.

In **adaptive direct synthesis control (ADSC),** added sophistication in the synthetic feedback program allows the DSC model of the system to be adjusted when actual feedback is received. ADSC controllers are used in modern machine tools to attempt to avoid anticipated problems (such as vibration at the tool), and to use sensory data after an output change to "learn" improved problem-avoidance response.

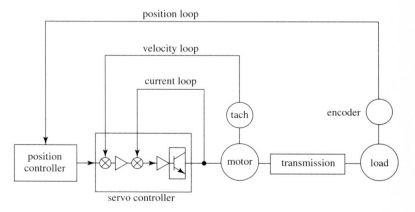

Fig. 6.27 Nested servo loops. (Diagram by permission, Allen-Bradley Canada Ltd., A Rockwell International Company.)

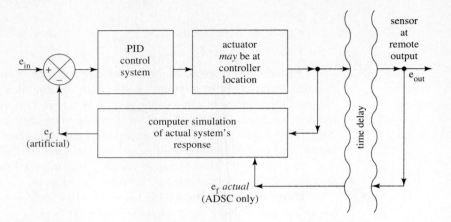

Fig. 6.28 Direct synthesis control and adaptive direct synthesis control

Systems described as **feedforward** systems are systems where the setpoints come from, or are modified by, sensed conditions *upstream* of the process being controlled. Feedforward controllers are sometimes called **cascade controllers.** The term *cascade* suggests that a change in an upstream variable will affect, or cascade down to, latter control systems. In its simplest implementation, the (amplified) signal from a sensor that measures incoming conditions might be used as a setpoint for another control system. A setpoint and a sensor's feedforward signal might be the inputs for an error amplifier that outputs a setpoint to another control loop.

The system in figure 6.29 is used to control the mixing of soda pop with artificial coloring. The first controller is a feedforward controller that receives a "final mix color" setpoint, subtracts sensory information about the concentration of the colorant on its way to the mixing tank, and outputs an "amount of colorant" setpoint to the next controller. The second controller, a standard closed loop feedback controller, controls flow of colorant to the mixing tank. The feedforward control system is predictive because it corrects out-of-tolerance conditions before they actually occur. It could be called cascade, because it uses

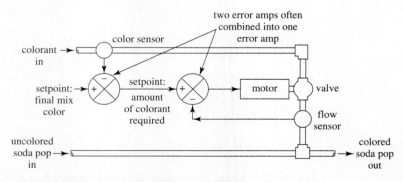

Fig. 6.29 A cascade feedforward control application

an open loop, feedforward, "final mix color" control system to change the "amount of colorant" setpoint that is provided to the closed loop, colorant flow control loop.

If a feedforward signal comes from a sensor measuring environmental conditions, it is called a **disturbance signal.** If the signal is provided manually, it is called an **offset signal.**

A **ratio controller** is a variation on feedforward control. It could actually be called a "feedsideways" controller. Its purpose, as demonstrated by figure 6.30, is to *use a sensor in one stream to control the processes in one or more parallel streams.*

In figure 6.30, the ratio controller's sensor measures the concentration of colorant, but not to control the flow of the colorant. Instead, the sensor provides a setpoint to two parallel fluid streams (soda water and flavor), so that the output from the mixing tank is the correct ratio of color, flavor, and soda water.

In **multiple output control** the output signal from a single controller is provided to *a group of parallel actuators*, then the result of the multiple actuation is sensed and fed back to the controller. The advantage of this type of control is that individual actuators can be adjusted separately, or even removed, yet the system's output will remain unchanged. Figure 6.31, p. 186, shows a multiple output controller causing a group of hydraulic cylinders to hold a load at a required height from the surface of a table.

In **constraint control,** more than one analog controller (or digital controller program) exists. Under normal operating conditions, one of the controllers does the controlling, but if a preset limit is exceeded at a sensor, the *second controller overrides the first* and takes over control. This type of control is easier to implement using digital controllers.

If a system controls, for example, the temperature in a beer-brewing vat, an integral controller would be ideal. When the same vat is first brought to the right temperature, however, the integral term is undesirable because it will build up and cause a large overshoot. Constraint control might be used so that while temperature error is large, the integrator is turned off and proportional control is used instead. When error is reduced to near zero, the integrator gain is turned high and the proportional gain is set to zero.

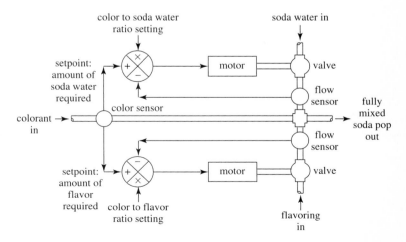

Fig. 6.30 Ratio control in a soda pop mixing process

6.6 DIGITAL CONTROLLERS

This chapter has primarily dealt with servocontrol theory as if the controllers were analog devices. In fact, modern servocontrollers and process controllers are now usually digital computers running control programs. The control programs are intended to make digital controllers behave like analog controllers.

A significant difference between digital and analog controllers lies in the techniques used to "build" them. Programming a digital controller means typing in a program containing control formulas. The program must input and output values via standard interface boards that the user adds in the computer. Tuning the servosystem requires typing in new gain factors. Changing the control algorithm means modifying the program. Copying the program can be done with a few keystrokes.

To program an analog computer, analog components must be wired together. Tuning is done by adjusting potentiometers or other adjustable circuit components. Altering the control algorithm means tearing down the controller and starting again. Copying a successful analog program means building a duplicate controller.

For true "real time control," an analog controller is required. The term *real time* means that a sensed change is reacted to immediately. Custom-designed, single-purpose analog controllers do this, although there is some deadtime due to electrical and mechanical energy losses. Digital controllers cannot operate in true real time, because they can execute only one program instruction at a time. There are at least 5 steps that inherently separate the time when a setpoint or feedback value is changed until the time when the response occurs:

1. Read setpoint
2. Read feedback
3. Subtract feedback from setpoint to find error
4. Multiply error by proportional constant
5. Output result

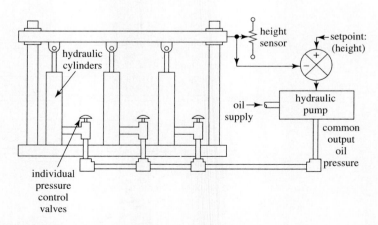

Fig. 6.31 Multiple output control

Luckily, with modern high speed microprocessors, program execution deadtime is insignificant for any but the most critical application. Most traditional suppliers of analog process controllers are now switching to digital controllers.

A look inside the cases of many robot controllers will show the reader that, while the main controller is a digital computer, each axis often still has its own analog controller. The analog controllers respond to setpoint changes provided by the digital computer. They correct disturbance errors without the digital computer's assistance.

For flexibility, reliability, and for the ease with which complex control algorithms can be generated and tested, analog computers cannot match digital computers. At the *upper end of the digital controller scale*, much research work is being done to find ways for **digital computer controllers** (for equipment such as robots) to solve increasingly **complex motion equations** that provide setpoints to servocontrol programs. Robot controllers offering controlled path motion are becoming cheaper and faster. With the growth in our ability to break up mathematical tasks so that a single equation can be processed by several microprocessors working in parallel, speed is being increased even further.

At the *other end of the digital controller scale,* many **programmable logic controllers (PLCs),** now offer built-in PID control at a low cost. I/O modules for these PLCs accept analog sensory inputs and output analog voltage or current to drive analog actuators. Coupled with a personal computer for online program entry, these PLCs offer the user the ability to select controller types (on/off, PE, PI, PID), adjust controller characteristics (gains), monitor the results, and even to select control configurations in software (feedback or feedforward, constraint control, direct synthesis, etc.).

One drawback to PLC servocontrol is speed. PLCs complete a program cycle in (typically) 20 milliseconds. This means that changes in the feedback may not be responded to for as long as 40 milliseconds (two program cycles). For many high and even moderate speed control applications, this is unacceptable. Some improvement is possible with PLCs that allow immediate data input and output during program execution. Several PLC suppliers offer servocontrol modules with quicker response characteristics, but which operate under the control of the PLC's main computer.

In the middle of the digital controller scale, **small-scale, microprocessor-based digital process control modules,** such as those shown in figures 6.32 and 6.33, are now offered by many suppliers. These controllers come complete with the terminations required to connect a wide variety of sensors and actuators, and have frontplate displays and controls that simplify the setting up of control functions. The user can adjust proportional, integral, and derivative gains and watch the input signal and the output signal displays simultaneously. Manual overrides are usually available, and these simplify setting up a system before switching it over to automatic control. Digital process controllers can control one, a small number, or many servosystems simultaneously.

Several digital process control modules are now available with **"auto-tuning"** features, relieving the user from having to set gain values. Some process control modules are now offering **"fuzzy logic"** (artificial intelligence) auto-tuning features. These controllers "think ahead" to work out how best to control their output *now* so that performance will be optimized in the *near future*. They can modify the way they control output as conditions change.

Suppliers of **large-scale, special-purpose digital process controllers** supply **development programs** with the equipment. The user uses the development program to configure the type of control and select gain values. Some of these development system programs allow the user to "build" an imaginary analog controller for the digital computer to control. All the user has to do is to start the development program from a keyboard, and select from menus the type of "components" to be programmed in. If the "circuit" doesn't work, a simple change at the keyboard alters it. Final implementation involves connecting sensors and actuators to the special-purpose input and output interfaces.

Recently, suppliers have been offering **process control software** that will run on personal computers. This software offers the same features available in other dedicated microprocessor-based process controllers. With this software, and the appropriate plug-in interface cards, any personal computer can be used as a process controller.

The **quirks of software-driven digital control,** and the quirks in converting digital signals to analog and vice versa, can cause unusual results if one is not careful. Digital computers can perform only one task at a time (because they generally have only one processor), and must perform that task in the order in which they are programmed to. This means that *output from sensors is examined only at periodic intervals, perhaps missing some important information*. It is entirely possible that the sensor sampling interval could **"bias"** the results. (For example, a computer might consistently sample an oscillating input just as the input crosses zero. The computer is led to believe that the signal from the sensor is always zero. Biasing was discussed in chapter 5, Vision Systems.)

Some digital-to-analog and analog-to-digital converters output short duration **erroneous signals.** These spikes can result in poor control if, for example, a digital controller samples a digitized input value while it is in error. Other sampling errors and quantization errors, inherent in converting between analog and digital values, are discussed in chapter 5. These dangers are well understood by manufacturers of commercially available servocontrollers, who know how to overcome them, but a user putting together a single customdesigned system may encounter difficulties.

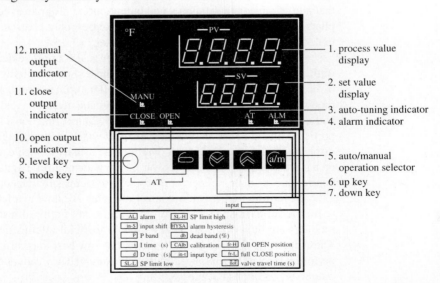

Fig. 6.32 An "off-the-shelf" digital temperature controller. (By permission, OMRON Canada Inc., Scarborough, Ontario, Canada.)

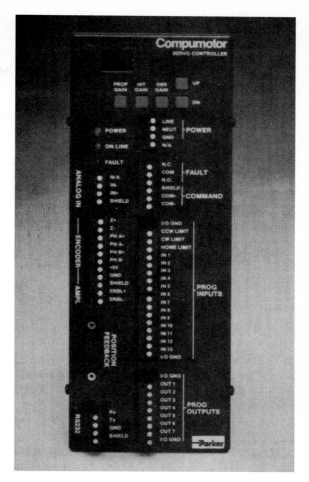

Fig. 6.33 A digital servomotor controller. (Photograph by permission, Parker Hannifin Corporation, Compumotor Division, Rohnert Park, California.)

Digital computers can be set up to respond faster by interrupting their current task when an input circuit detects an important condition. While the use of **"interrupt-driven"** programming can prevent the "biasing" problem previously discussed, it can lead to new problems if the task the computer is performing at the time of the interrupt is also an important task.

6.7 SUMMARY

In this chapter, we have concentrated on **closed loop,** or **"servo" control.**

We have seen that **transfer functions** can be used to describe the behavior of a component or of a complete control system. Graphical techniques, including **Bode diagrams** and **Nyquist plots,** for depicting system or component behavior have been examined.

We discussed system **gain, phase lag, resonance,** and **instability** as observable using graphical techniques. We also used graphical depictions of system response to examine some other servosystem response characteristics, such as **overshoot, steady state error, deadband, hysteresis, saturation, deadtime,** and **capacitive time constant.** We observed that the response of a servosystem is, in part, dependent on the type of **input signal.**

We examined several types of control systems, from simple on/off controllers to the current state-of-the-art **PID controllers,** and examined the characteristics of the proportional, integral, and derivative terms in the PID controller.

We examined some variations on the ways in which control systems can be used other than as simple single loop controllers. We saw variations such as **nested loop, feedforward, synthetic feedback,** and **constraint control.**

Finally, we discussed some of the various types of controllers available to the system designer, from the traditional **analog controls,** through **digital controllers** fitted with faceplates that make them look like analog controllers, to **PLC modules** that perform PID control under PLC control.

REVIEW QUESTIONS

1. In what way is closed loop control better than open loop control?
2. a) What does an error amplifier do in a servosystem?
   b) How is the error signal used in a servosystem?
3. What is meant by the term "G" in a servosystem block diagram?
4. What does a servosystem transfer function describe?
5. Identify three methods of graphically describing a servosystem's behavior. Briefly describe what each of these graphical forms depicts.
6. Why might a servosystem exhibit gain loss at high frequencies of error signal?
7. Identify three possible causes of resonance in a servosystem responding to a repeating input signal.
8. Sketch a position/time graph for the following position control servosystems:
   a) a stable, underdamped system.
   b) a system with monotonic instability.
   c) a system with oscillating instability.
9. Identify two possible sources of disturbances requiring servosystem response.
10. a) How does an increase in friction affect system damping? How does an increase in inertia affect damping?
    b) How does increased friction affect acceleration of a DC motor driven servosystem? An increase in inertia?
11. In observing the response of a servosystem, deadband or deadtime might be observed. Define deadband and deadtime and identify one possible source of each.

12. If a controlled system is described as having a time constant of 0.2 seconds, how close will the system's output be to its final position 0.4 seconds after a unit step setpoint change (assuming no deadtime)?
13. How does differential gap control differ from simple on/off control?
14. a) What two types of error are typical in the response of a proportional controlled system?

    b) A PID controller should correct both error types. Identify which term in the PID equation (pick any PID equation) is intended to correct each error type. Describe how that term in the equation corrects that error type.
15. How is proportional band related to the saturation levels of an actuator in a control system?
16. Define "repeats per minute" as it applies to integral control.
17. The designer of a derivative control system can choose to use the derivative of either of two values. What are the two values? For each value, identify one possible advantage over the other.
18. a) What three general factors affect the response of a controlled system?

    b) You are asked to troubleshoot an underdamped control servosystem. The control system is supposed to respond to a series of computer-fed setpoints by moving a machine bed to new positions and holding it there until the next setpoint is issued. The machine bed moves to the new positions quickly, but tends to take a long time to settle after the initial overshoot. Suggest one specific example solution from each category from part (a) of this question, and describe how that change would improve system damping.
19. Under what circumstances might direct synthesis control be used in place of true feedback? (Give an example.)
20. In what way might a control system with feedforward be better than one without feedforward?
21. Even very sophisticated robot controllers often have analog servocontrol circuits for positioning each individual axis. The digital controller supplies the analog controller with a series of position and speed setpoints, but otherwise leaves position and speed control to the analog controller. Why have two different types of control?
22. You want to use a PLC for servocontrol of a fast-moving position control system. What would you look for in the PLC system?

CHAPTER 7

CONTROLLERS: DIGITAL COMPUTERS AND ANALOG DEVICES

CHAPTER OBJECTIVES

After completing this chapter, the reader will be able to:

▬▬▬ Discuss digital and analog controllers as used in industrial automation.

▬▬▬ Describe the components of a computer and of a PLC, and describe what these components do.

▬▬▬ Discuss the various levels of software running on a personal computer or a PLC during a typical session.

▬▬▬ Describe various components that allow computers to input, output, and exchange data, including:
- parallel I/O, opto-isolators, line drivers
- analog to digital and digital to analog conversion
- serial I/O, RS 232, RS 485, and similar standards
- modems
- broadband, baseband, and carrierband networks

▬▬▬ Distinguish among hierarchical, star, bus, and ring local area network architectures, and token-passing and CSMA/CD protocols, and discuss the merits of each.

▬▬▬ Discuss several silicon-based controller components, including:
- diodes and Zener diodes
- bipolar and field effect transistors
- op amps as comparators, error amplifiers, proportional amplifiers, integrators, and differentiators
- SCRs, GTOs, and triacs

▬▬▬ Discuss trends in analog and digital controls.

7.0 CONTROLLERS: DIGITAL COMPUTERS AND ANALOG DEVICES

The silicon revolution has changed manufacturing! There are "lights-out" factories, working non-stop, where humans are called in only when repairs are necessary. The world's largest manufacturers have installed Computer Integrated Manufacturing (CIM) systems to schedule, track, and even to perform production tasks. Some manufacturers have their computers connected via Wide Area Networks so that they receive daily data updates from other corporate facilities worldwide (e.g., product design changes). Even tiny manufacturers find they need some automation, in the office and on the production floor.

7.1 LEVELS OF CONTROL AND MAJOR COMPONENTS

For a true "intelligent" and flexible automated manufacturing system, the system should be able to respond to changing conditions by changing the manufacturing process all by itself. The system should be reprogrammable so that the equipment can be reused when the product demanded by consumers changes. Digital computers can be programmed to "read" external conditions and to execute different parts of their programs depending on the sensed conditions. Digital computer programs can be written and changed easily. There are even computer programs to write and run other computer programs!

Figure 7.1 shows computer-assisted steps in producing a metal part. A human designs a part on a Computer Aided Design (CAD) system, requests that the Computer Aided Manufacturing (CAM) software write the numerical control (NC) program, then sends the program to a computer-selected NC machine, which produces the part.

A computer-based control system need not be as powerful as that shown in figure 7.1. It is, in fact, not advised that a user attempt to install such a system until the user has experience with less sophisticated systems.

7.1.1 Workcell Peripherals

The focus of an industrial automation control system is at the workcell. Those devices that are monitored and controlled by computers are called **peripherals**. Peripherals may include:

- **Keyboards, monitors, printers,** or **diskdrives.** While these are essential to a personal computer, an industrial controller can often do its work without any of them.
- **Sensors and actuators.** To control an industrial process, the computer must "see" the process through its sensors, and change the process through its actuators. Whole chapters are devoted to sensors and actuators in this text.
- **Intelligent peripherals** such as remote sensors and actuators with communication adapters. A computer can be connected to several communication adapters with a single set of conductors, and can read selected sensors or write to selected actuators individually, specifying each by an address.
- **Other computerized controllers** such as numerical control (NC) equipment, robots, vision systems, and other computers.

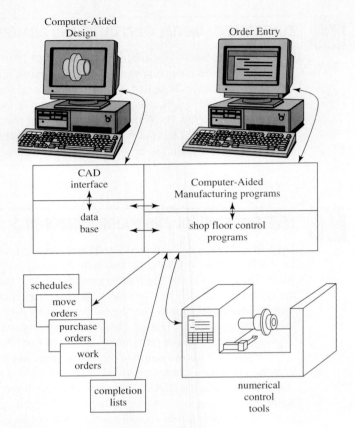

Fig. 7.1 A CAD/CAM part production system

7.1.2 Workcell Controller

If none of the computers already in the workcell are powerful enough to control the workcell's components, then a workcell controller is needed. An obvious choice for a computer to use as a cell controller is a standard **personal computer (PC).** Such a computer can, however, be difficult for the novice to use for automation control. A PC will require additional interface cards and programming. The setup person needs to be trained to select the hardware and software, to design the electrical connections, and to write the special-purpose programming.

If the required expertise is available, a PC-based system such as that shown in figure 7.2 can be assembled using purchased interface cards and software packages. The personal computer itself does not have to look like the standard office version. They can be purchased in "industrially hardened" versions with increased resistance to abuse and dirt, or purchased as single cards that can be assembled into user-built cases.

Most new users of automation control start with a **programmable logic controller (PLC).**

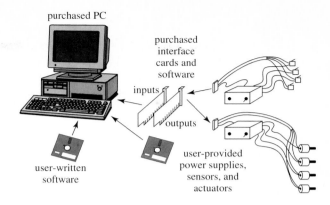

Fig. 7.2 A personal computer with peripherals for a control application

Even the simplest PLC comes complete with interface hardware, programming software, and a wide array of easily-connected expansion modules available from the PLC manufacturer. Figure 7.3 shows that fewer user-selected components are required after PLC selection for the same application as that shown in figure 7.2. Here, the user need only select and add appropriate sensors and actuators, and write the program.

PLCs are usually programmed using a language known as **ladder logic,** which was developed to be easily understood by industrial technicians familiar with relay circuit design. The language (rather strange in appearance at first sight) is easily learned by people who have never programmed before, even if they are not familiar with relay circuits.

In the early 1980s it looked like PCs would easily replace PLCs on the plant floor. PLCs were handicapped by the limitations of early ladder logic. Communication between proprietary PLCs and other controllers was difficult. Early PCs needed only a few standard interface cards and user-friendly programming software packages to take over. The interface cards and software never materialized. Meanwhile, PLC suppliers expanded their offerings to allow networking of PLCs, improved the programming languages, and offered programming software so that PLC programs could be written at standard PCs. With the wide range of off-the-shelf interface modules available for PLCs, it became common to use a PLC as the main controller in a workcell.

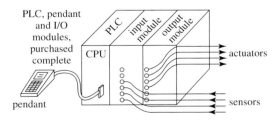

Fig. 7.3 A PLC as a cell controller

7.1   Levels of Control and Major Components   **195**

Mainstream computer manufacturers have had a belated awakening and are now offering what they identify as **"cell controllers."** Cell controllers are primarily intended to handle inter-controller communications, to act as data storage devices, and to provide operators with a central control panel from which whole manufacturing systems can be controlled. Suppliers of PLCs are also well-positioned to capture the cell controller market, sometimes with equipment developed in cooperation with the mainstream computer manufacturers.

7.1.3 Central Controller

Without a central controller coordinating the functions of the cells, each cell is simply an "island of automation." While the creation of islands of automation during a company's learning cycle is almost unavoidable, they should be created with an eye on the eventual need to interconnect them.

Central controllers are often powerful mainframe computers, and are usually called "host" computers. Microcomputers are becoming powerful enough to do the job in some cases. In figure 7.4, we see that the host computer maintains the **database.** The database consists of:

- **product data** (perhaps created on the CAD system)
- **financial data** (from marketing)
- data on the **manufacturing capabilities** (from manufacturing)
- **individual programs** for communicating with workcells (to download programs or to read production status)
- the status of **work in progress (WIP)**
- the current **stock levels**
- **outstanding orders**
- **projected completion dates**
- certain other **variables** (such as limits on stock levels, manufacturing strategy, etc.)

The host computer has a **database management system (DBMS)** program, which allows access to the database. DBMS commands can be used to read or write to the database, and to permit settings to limit the data that is available to individual users.

The plant data are used by a group of programs that together are known as **Manufacturing Resource Planning (MRP II)** programs. MRP II can be used to predict capability of the plant to meet predicted orders, to optimize the scheduling of production for maximum throughput, or to schedule so that work is completed on time. MRP II programs output detailed production schedules and work orders for individual work centers, and accept input to modify the schedules as required.

The central computer also contains programs to **control communications** on the network. Networking is discussed in the communication section later in this chapter. Chapter 9 discusses the role of a central controller in automated manufacturing using examples.

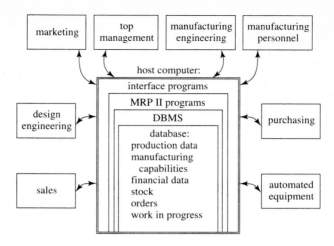

Fig. 7.4 Central controller: database, programming, and networking

7.2 DIGITAL COMPUTERS, THE HEART OF AUTOMATED CONTROL

In the following sections, we will examine how a computer works and what types of external computer components are required to monitor and control an automated process.

7.2.1 Computer Architecture

A computer consists of three types of devices, interconnected by three sets of conductors called "buses." Figure 7.5 shows:

- central processor
- memory devices
- input/output devices

all interconnected via:

- a data bus
- an address bus
- a control bus

 The **Central Processing Unit (CPU)** is the unit that controls the computer. The CPU reads the computer program from memory and executes the instructions. It manipulates values stored in memory and inputs or outputs data via input/output (I/O) devices. The CPU will be discussed in more detail later in this section.

 The buses are sets of parallel conductors. An 8 bit computer has 8 conductors in its **data bus** and thus can move 8 bits of data at a time. A 16 (32, or 64) bit computer has 16 (32, or 64) data bus conductors and can process 16 (32, or 64) bit data "words" at a time. Some computers use bus-sharing techniques to reduce the number of conductors required in the data bus.

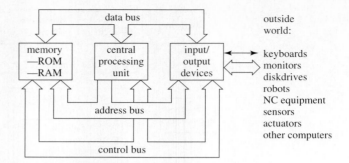

Fig. 7.5 Basic computer architecture

Each 8 bit memory location and I/O port has a unique address. The CPU uses the **address bus** to identify which of the memory locations or I/O locations it is about to read from or write to. Each conductor in the address bus can carry only either a 1 or a 0, so there is a limited number of different addresses that can be specified.

The **control bus** contains miscellaneous other essential conductors. Some conductors provide the 5 volt supply and ground lines. Others are controllable by the computer components. The CPU, for example, uses the read/write line in the control bus to specify whether it intends to read from or write to the address that it has put on the address bus. There are several "interrupt" lines in the control bus that normally carry 5 volts and that, when dropped to 0 volts, cause the CPU to respond immediately even if a program is already running. Interrupts are used to initiate a computer response to a situation that cannot wait, such as a fire in the warehouse, a computer reset, or simply a keyboard entry (which must be read before the human presses another key).

Some computer systems include a memory/IO line in the control bus, thereby as much as doubling the number of addresses that the computer can specify. Only half of the addresses can be in memory, while the other half must be I/O ports.

Memory in figure 7.5 refers to **internal memory.** Programs on a disk or other mass storage device must be read into this memory via an I/O port before the CPU can run the program. There are basically two types of internal memory: random access and read only.

Random Access Memory (RAM) can be written to and read from by the CPU. RAM memory is lost when computer power goes off, unless a backup battery maintains power to the memory. Many industrial controllers (such as PLCs and robot and NC controllers) have battery-backed RAM, so that user-programs do not have to be reloaded into memory every time the controller is powered up.

Read Only Memory (ROM) can only be read by the CPU, and is not lost when the computer is turned off or unplugged. ROM chips come in several types. Some ROM chips are **custom manufactured** where large volumes warrant. **PROM** chips can be programmed once by the user, using a PROM programming device. They are used by medium-quantity suppliers if technology changes fast enough to rule out the cost of tooling-up for custom ROM (e.g., a robot supplier). Erasable PROMs (**EPROM**) can be programmed like PROMs, but can be erased for reuse by exposure to high-intensity ultraviolet light. They are used by computer suppliers who write their own programs for storage in ROM, but who make frequent changes to those programs.

Electrically Erasable PROMs (**EEPROM** or **EAPROM**) are commonly supplied with PLCs. They are used very much like floppy disks, but they do not require expensive, fragile read/write mechanisms. User-programs can be copied from PLC RAM onto an EEPROM chip in a plastic carrier, then the EEPROM can be removed for use at another PLC, or can be left in place for copying back into RAM at startup.

Flash memory, often on removable **"PCMCIA"** (Personal Computer Memory Card International Association) cards, is now available. The computer reads and writes flash memory as if it was RAM, but data is not lost when power is removed.

7.2.2 Computer Sizes

Miniaturization of integrated circuits has reached the point where the whole computer can be printed onto a single chip! If a single-chip computer is manufactured complete with pre-programmed ROM, and with RAM and I/O ports specially suited for a certain task, it is called an **Application Specific Integrated Circuit (ASIC).**

General-purpose single-chip computers with PROM, with significant RAM, and with I/O ports that can be configured for user-selected purposes are known as **Micro Controller Units (MCU).**

If the chip includes only the CPU, it is called a **Micro Processor Unit (MPU).** An MPU is typically combined with memory and I/O chips on a single circuit card, perhaps with slots for user-added expansion cards, and is then sold as a **microcomputer** or as a **personal computer.** Microcomputers often come with slots for special-purpose slave processing units called **co-processors.** Math co-processors are available to enhance an MPU's number-crunching speed. Graphics co-processors are available to speed the rate at which high-resolution video displays can be presented.

Minicomputers contain several cards, and often have more than one CPU chip. Powerful new microcomputers are making minicomputers obsolete.

Mainframe computers often come with several cabinets full of components comprising the single computer. **Supercomputers,** like mainframes, are often large enough to occupy several cabinets. They are even more powerful and faster than mainframes.

7.2.3 The CPU

A CPU contains several types of **registers** in which binary numbers can be stored. Figure 7.6 shows the different types of registers typically present.

Every CPU or MPU has at least one **accumulator,** where it holds the data it is using. With some exceptions, the CPU can change data only if the data is in an accumulator.

To keep track of where it is in a program, the CPU stores the memory location of the next instruction in a register called a **program counter.**

Index registers are used to store address offsets. Whole blocks of data (even programs) can be stored anywhere in RAM, with their starting addresses stored in index registers. Programs that use the data also use the index register to determine where the block of data starts. Index registers make it possible for a program to be loaded into memory and run without first clearing other programs or data out of memory. Most CPUs contain several index registers. Programs can use unused index registers for counting.

The **condition code register** keeps track of conditions in the CPU, using the individual bits of the register as "flags." As an example, if two 16 bit numbers are added in a

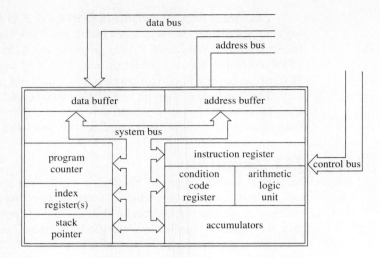

Fig. 7.6 CPU registers

16 bit computer and the sum is a 17 bit number, a carry flag bit in the condition code register is set. Flag bits are checked by conditional branching instructions. Depending on the condition of the checked bit(s), a branching instruction may change the address in the program counter, so that program flow is changed.

Another condition code register bit is an **interrupt mask bit.** If the program that is running is very high priority, it will include an instruction to set this interrupt mask bit. When the voltage drops on an interrupt line in the control bus, the CPU will refuse to respond to the interrupt while the interrupt mask bit is set. (Some interrupt lines in the control bus cannot be masked by the interrupt mask bit.)

The accumulator(s), index register(s), program counter, and condition code register are accessible by the user program. This means that program instructions can change the contents of these registers.

A CPU contains **other registers and buffers** that are not specifically changeable by any instructions and that will not be discussed here. The stack pointer, the instruction register, the address buffer, and the data buffer fall into this second category. Also not discussed here is the arithmetic logic unit (ALU), essential to the CPU's operation, but that can be taken for granted by the user. The next section demonstrates how the CPU registers are used in a program execution.

7.2.4 Operating Systems, Software, and Memory Organization

The CPU, even if it is the heart of a sophisticated CIM system, is really only an electric circuit. It contains large numbers of transistors, so it can be easily changed from a circuit with one purpose, to a circuit with another purpose. When power is off, all those transistors are off, so the circuit always starts the same.

This section will follow the operation of a personal computer's CPU circuit as it "wakes up" and starts to run a series of programs. Please refer to the computer and MPU architecture diagrams in figures 7.5 and 7.6, to the memory map in figure 7.7, and to the structured flowchart of operations in figure 7.8 during reading of this section. The instructions mentioned in this section are binary machine language instructions. Later, we will discuss how machine language programs are created from higher level language program listings.

1) At startup (or reset), the initial CPU circuit causes the CPU to read the contents of the last memory location into the program counter.

This last memory location, which must be a ROM memory location, is pre-programmed with the address of the first instruction of the first program that the computer is to run, and is known as the **reset vector.**

2) The contents of the program counter are then copied into the address buffer, and therefore onto the address bus. The address in the program counter is incremented to point to the next memory location, ready for the next operation. The read/write line of the control bus is set to "read."

The indicated memory location places its contents onto the data bus. This binary data moves into the data buffer, then into the instruction register. The new contents of the instruction register dictate what the CPU circuitry is to do next. This data word is the first instruction of the program that is often called the **Basic Input/Output System (BIOS).**

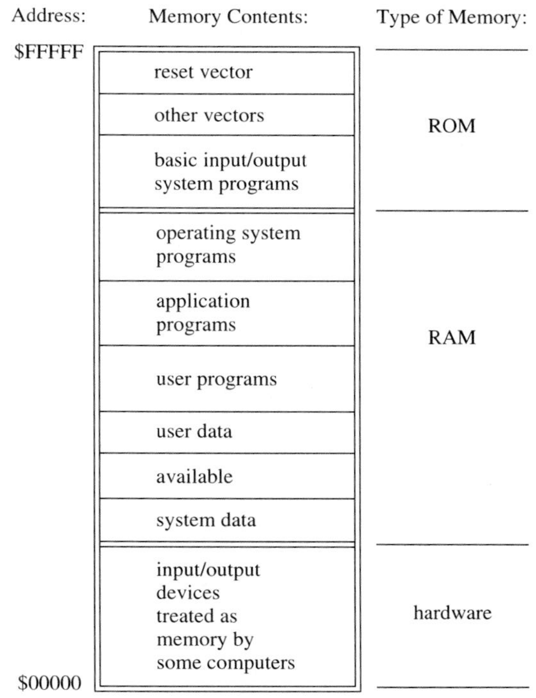

Address:	Memory Contents:	Type of Memory:
$FFFFF	reset vector	ROM
	other vectors	
	basic input/output system programs	
	operating system programs	RAM
	application programs	
	user programs	
	user data	
	available	
	system data	
$00000	input/output devices treated as memory by some computers	hardware

Fig. 7.7 Memory map

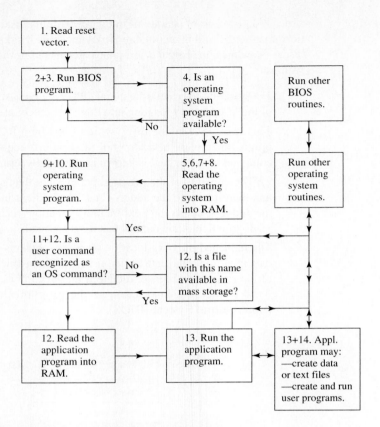

Fig. 7.8 Flowchart of CPU operations from startup

3) Once this first BIOS instruction has been executed, the next instruction is read from the next memory location. The BIOS program runs one instruction at a time, in this manner. The Basic Input/Output System instructions must be in ROM, because they tell the computer how to be a computer, and must be available every time the computer starts up.

The Basic Input/Output System program typically causes the computer to run a self-check routine at startup. It also includes routines that can later be called by other programs to use the miscellaneous I/O devices present in the computer (e.g., to read or write to a disk drive).

4) Once the self-check is complete, the Basic Input/Output System will check for the presence of a mass storage device, such as a floppy disk or an EEPROM. If a mass storage device is present, the BIOS assumes that it holds a more sophisticated **Operating System (OS)** program (such as MS-DOS®, UNIX™, OS/2®, or WINDOWS NT™).

5) If present, the operating system program is then read into RAM memory via an input port as follows: The Basic Input/Output System writes to an output port,

telling the mass storage device to start sending the operating system program to an input port of the computer. When the input port has received the first full data word from the floppy drive system or EEPROM, it notifies the CPU via an interrupt line.

6). This first data word often indicates the number of data words contained in the operating system program. The CPU reads this number from the input port and puts it into an index register. (Call this index register "X.") Meanwhile, the input port is receiving the next data word from the floppy disk.

7) The next incoming data word is read into the CPU, then stored in an unused RAM memory location indicated by a second index register (call it "Y"). Index register Y will be incremented as each data word is read into RAM, so that subsequent write-to-memory operations will not write on top of data already read in and stored.

8) Index Register X is decremented. If it has not yet been decremented to zero, step 7 is repeated.

9) Now that the complete operating system has been read into RAM, the program counter is loaded with the address of the first RAM memory location of the operating system program. The first operating system instruction is then read into the instruction register, and the operating system program takes control of the computer.

10) The operating system program will eventually print a prompt (such as "C:\>") on a monitor, then wait. This program will allow the user to give commands to the computer.

11) When an operator presses any key at the keyboard, an interrupt is generated, which causes the operating system program to read the key value and store it in system data RAM.

12) Eventually, the operator's keystroke will be an [Enter]. When the operating system recognizes an [Enter], it examines the RAM locations holding the complete operator's keyboard entry.

If it can recognize the command as an operating system command (such as dir, copy, load, etc.), it will branch to the portion of the operating system program that can follow the order.

If the keyboard entry is not an operating system command, the operating system will often search for a file with that name (e.g., Lotus®, WordPerfect®, BASIC, etc.) on the mass storage device. If the file is found, the operating system will use the Basic Input/Output System programs to read the external program into unused RAM, then jump to the start of that new program.

We will assume that a non-operating system command has caused another program to be read into RAM, sharing RAM with the operating system.

13) The new program, often called an **application program,** is a special-purpose program that performs some function such as word processing, computer programming, data communication, or manufacturing control. The application program can jump to portions of the operating system or the Basic Input/Output System, treating them as subroutines, and often can be interrupted by external events

such as keyboard entries. Some application programs can be used to create and run yet another level of program, the **user program.**

14) User programs are usually written in "high level" languages such as BASIC, Pascal, or C. In order for the computer to understand and run them, they must be translated into machine language. One of the functions of programming language application programs (such as BASIC, Pascal, or C) is to do this translation. We say that the high level language user programs are **interpreted** or **compiled** into machine language.

15) Eventually, each level of program reaches an end, returning control to the previous program level and releasing its allocated RAM.

Some controllers used *in industrial control* do not need to go through as many steps at startup as a personal computer has to. As single-purpose computers, their *operating system program and application program are pre-loaded into ROM memory* with their Basic Input/Output System programs. PLCs, numerical control machinery, and robot controllers are examples. Robots will be discussed in detail in a later chapter and numerical control is similar to robot control. The rather unusual (for a computer) behavior of programmable logic controllers will be examined in more detail in this chapter, after we have examined input/output devices.

7.2.5 Input/Output Devices

I/O devices (sometimes called "ports") are the computer components that the computer uses to exchange data with its peripherals and/or with other controllers.

Data inside a computer consists of binary data words, which are moved via data buses between CPU registers and other chips. Data outside a computer may also consist of binary bit patterns on *parallel* conductors, may consist of binary bit patterns carried one at a time on a *single* conductor pair, or may be *analog* DC or AC signals on a conductor pair. The function of an I/O device is to translate data between the form required inside the computer and the form required outside the computer. This function is often called **interfacing,** or sometimes **signal conditioning.** Figure 7.9 graphically demonstrates the various changes in signal type that may be necessary for a local digital controller to use input from sensors, provide output via actuators, or to exchange data with other controllers via a Local Area Network in a Computer Integrated Manufacturing system.

Before examining I/O signal conditioning devices, a short review of the terminology to be used is in order.

Digital (or "discrete") means that there are only a fixed number of signal levels possible. The term *digital* implies only two levels. *Discrete* implies a limited number of possible levels on one line, or a limited number of possible combinations of digital signals on several lines.

Digital level usually implies voltage (0 to 5 V) and current levels compatible with a computer. Digital level is also called **"TTL level."**

Analog (or "continuous") means that any signal level is possible, within the minimum to maximum range (or "span" or "bandwidth") of the signal source.

Configure means to set up an I/O device for a specific function. Some I/O devices are pre-configured. Pre-configured I/O devices are appropriate for common peripheral

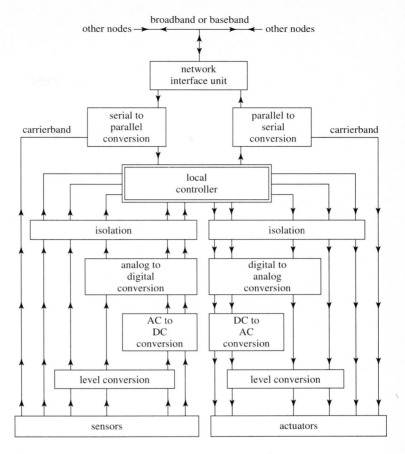

Fig. 7.9 I/O interfacing requirements of a local controller

devices such as keyboards, monitors, etc., but if the user intends to connect peripherals such as sensors, motors, etc., it is often best to use programmable I/O ports.

We will now look at several different types of I/O devices.

Parallel I/O

For fast data input and output, whole data words can be sent or received, with all their bits carried simultaneously on parallel conductors. Data sent to a parallel output port by the CPU is "buffered" in the output port (stays in the most recently written state), until the CPU writes to it again.

Sensors, actuators, or other computer peripherals such as printers can be connected to a computer via parallel I/O ports. Computers do not usually communicate with other computers via parallel interfaces.

If the peripheral is **an intelligent peripheral** (one with a built in computer controller), the whole exchange of parallel data can be at **digital level,** so the I/O device may not need to do anything more than buffer the data words. For long distance data exchanges ("long" might mean anything over 2 feet), **signal isolation** may be necessary. **Line drivers** may be needed to provide stronger signals.

Where two electrical circuits are near each other, capacitive and inductive effects (not to mention short circuits) can lead to changes in both. In electronic circuits, the signals unintentionally generated in one circuit due to changes in a nearby circuit are known as "crosstalk," and except in the world of spying, are considered undesirable.

Signal isolation via an **opto-isolator** chip is shown in figure 7.10 (a). Each parallel line from the parallel output buffer chip controls an LED inside the opto-isolator chip. The light from each LED strikes a phototransistor, which allows current to flow in another circuit when light is present. There are no electrical connections between the input and output circuits, so there is no possibility of electrical noise in one circuit damaging the other circuit. When rated for power levels above digital level, opto-isolators are sometimes called **Solid State Switches,** or sometimes **Optical Switches.** (The output circuits in the diagram are shown at 24 volts to emphasize that they are isolated from the internal 5 volt circuits.)

Line driver chips allow low power circuits to switch higher power circuits. Using op amps instead of opto-isolators as switches, the input and output circuits are not as well isolated from each other. Figure 7.10 (b) shows how two parallel I/O devices can be connected via line driver chips. The diagram demonstrates that line drivers can be used to allow the computer to control a circuit with higher voltage levels. One advantage of using

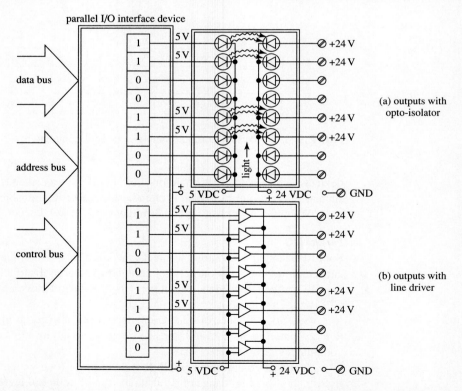

Fig. 7.10 Parallel I/O connections with (a) opto-isolation and (b) line driver chip

op amps is that they can be used to maintain current levels in an external circuit despite variations in the external circuit's resistance.

The peripherals connected at a parallel I/O device may be digital **actuators and switches.** Each switch or actuator must be assigned its own line.

Switches and some small actuators (such as LEDs) can be connected at **digital level** as shown in figure 7.11 (a). Switches, even operating at digital levels, must be connected in such a way as to allow charge to drain away from the I/O device when the switch is closed.

Signal isolation and amplification are often required. As shown in figure 7.11 (b), **opto-isolator modules** are available, and can be purchased to even interface high current level AC to digital level signals. **Relays** can also be used, but are largely being replaced by opto-isolator modules.

Some sensors have very weak outputs, and should be interfaced to the computer digital I/O device via **operational amplifier (op amp)** chips (see figure 7.11 (c)). The advantage in using an op amp is that the weak signal from the sensor is not reduced by driving the op amp.

Configuring of parallel ports consists, partly, of programming the I/O chip to use some lines as inputs and others as outputs. Other configuration options include whether to allow the I/O chip to interrupt the CPU when it receives changed input values.

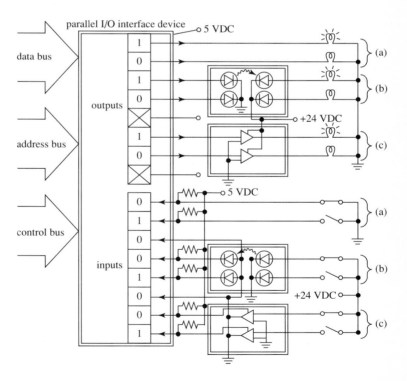

Fig. 7.11 Digital actuators and switches connected at digital I/O devices: (a) without any other signal conditioning; (b) with opto-isolator modules; (c) with operational amplifiers.

Analog I/O

Often, a computer must "read" the value of an analog voltage or current, or must output an analog value. Analog I/O signal conditioning devices are needed.

In the Motors chapter, we discussed how motor controllers could vary DC voltage level and AC frequency. The methods discussed were those by which a computer could output analog signals. They included PWM and VVI.

Pulse Width Modulation (PWM) employs power from a large DC power supply that is switched on and off by a digital controller. The resultant **analog DC voltage output** depends on the percentage of time that the power is switched on. For **AC frequency** control, DC is switched between positive and negative. The resultant voltage at any time is the average of the recent DC voltage levels. The average voltage value is varied so that it rises and falls at the selected frequency. A computer can be programmed to do this switching via a single parallel I/O device circuit, or can control a PWM chip that does the switching.

Variable Voltage Inversion (VVI) is used for controlling AC. VVI involves switching two positive and two negative DC supplies to yield approximate AC at a selected frequency. A computer can switch four non-digital level circuits using four channels of a parallel I/O device.

In addition to these methods of controlling analog I/O, industrial control computers must often output small analog DC levels without using switching techniques. They also must receive analog values and convert them to digital data words.

Digital to Analog Converters (DACs, or D/A converters) convert digital numbers to analog DC signals. The most common DAC is a single chip containing an **R-2R** circuit, as shown in figure 7.12. The analog value outputted is proportional to the value of the digital number the CPU writes to the DAC chip. There are obviously only a limited number of different analog values possible (the number of different digital numbers possible), so the output is not truly "analog," but is instead "discrete." The small analog DC output of a DAC must often be further amplified.

If a sensor's analog DC output must be read by a computer, the signal must be converted into a digital-level binary number. Dedicated IC chips, known as **Analog to Digital Converters (A/D converters or ADCs),** are available for this function. In general, ADCs compare the input analog DC signal against a reference voltage, or a series of internally-generated voltages, to decide its approximate value. ADCs are selected for their cost,

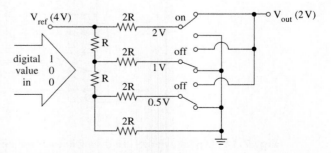

Fig. 7.12 The R-2R digital to analog conversion circuit

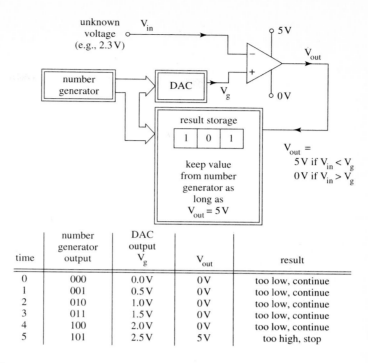

Fig. 7.13 Ramp analog to digital conversion

speed, and resolution. The more bits in the binary number it outputs, the greater the resolution of the ADC.

The simplest ADC uses the **ramp** method. As demonstrated in figure 7.13, the ramp ADC generates a series of increasing binary numbers, each of which is converted to an analog voltage level (by a built-in DAC). These voltage "guesses" are supplied to a comparator op amp. The unknown voltage is connected at the other op amp input. Eventually, a "guess" voltage will exceed the unknown voltage, and the most recently generated binary number will be used to represent the analog voltage. This process takes longer if the unknown voltage is at a high level, and of course, takes longer if high precision requires many guesses to be made. Ramp ADCs are not often used.

Perhaps the most widely-used ADC is the **successive approximation** type. This ADC, demonstrated in figure 7.14, works like the ramp ADC, but the "guess" voltages start at the middle of the voltage range. If the comparator indicates that the unknown voltage is higher than the first guess, then the second guess is midway between the first and the maximum. Using successive guesses in this manner, the ADC zeros in on unknown voltage. It needs one guess for each bit of its binary output word. Conversion always takes the same amount of time.

Dual ramp ADCs are commonly used, also. See figure 7.15. In this type of ADC, the unknown voltage is allowed to charge a capacitor, through a resistor, for a fixed time. After this time, the unknown voltage is disconnected, and the capacitor and resistor are then connected to a reference voltage with opposite polarity. The time to discharge the

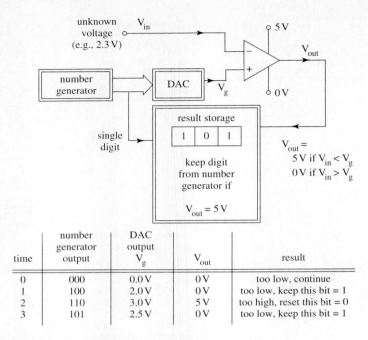

Fig. 7.14 Successive approximation analog to digital conversion

capacitor is clocked. The discharge time, of course, depends on how much charge the unknown voltage moved into the capacitor. The binary number representing the discharge time is the binary representation of the unknown voltage. The process is somewhat immune to noise, because the capacitor effectively averages the charging voltage for a short period. Dual ramp ADCs compensate for fluctuations in the supply voltage by also using the sensor's supply voltage as the reference voltage.

The **flash converter,** shown in figure 7.16, is very fast. It consists of a large number of op amp comparators, each supplied with a different reference voltage. The unknown voltage is connected to all comparators simultaneously. A binary number, depicting the number of comparators at which the unknown exceeds the references, is generated. This large and expensive ADC is used where large amounts of data must be captured quickly, such as in video image capturing for industrial vision.

The recently-developed **economical ADC** is a "flash" variation on the successive-approximation ADC. A simplified 3-bit output economical ADC is shown in figure 7.17. One half the reference voltage is compared to the unknown in the first comparator, which outputs if the unknown is higher than its (halved) reference. Each of the next comparators' references are half of the reference of the previous comparator, increased by the value of the reference voltage of all *of the previous comparators that are outputting.* This ADC, therefore, instantly generates successive approximation reference voltage guesses to compare with the unknown, and requires only as many comparator circuits as the size of the digital output data word.

The **tracking converter** is also fast. In this converter, the input voltage is constantly compared against two reference voltages, one slightly above and one slightly

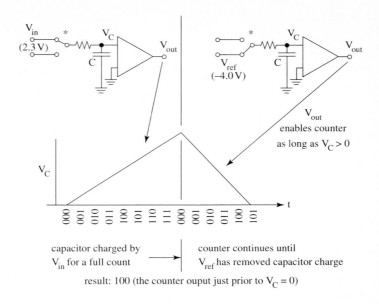

Fig. 7.15 Dual ramp analog to digital conversion

below the "now" input value. While the input remains between the reference values, the stored "now" value remains unchanged. If the input value wanders out of this range, the "now" value is incremented (or decremented). The reference voltages are also increased (or decreased). This converter actually does work only when the unknown DC voltage is changing, but responds immediately to that change. Tracking converters can be set up to

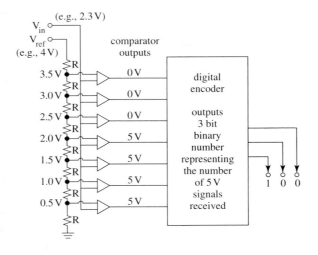

Fig. 7.16 The flash analog to digital converter

7.2 Digital Computers, the Heart of Automated Control   **211**

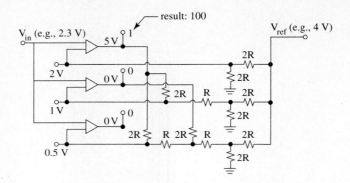

Fig. 7.17 The economical analog to digital converter

notify a controller when they have detected a new voltage level, saving computer time. Tracking converter error, unfortunately, is cumulative.

In a recently developed ADC, the unknown voltage creates a charge that deflects a **laser beam** (all inside a single silicon chip). The laser strikes one of several patterns of output sensors, depending on how much it is deflected. The sensor pattern is outputted as the digital value of the analog voltage.

If the analog signal to be input or output by the computer is an **AC signal,** other chips, using other techniques, are required.

AC signals may be sampled and *converted to digital* values at a rapid rate. The signal must be sampled frequently enough to get a good digital representation of the AC waveform and its peak values. The Vision chapter discusses sampling and potential errors.

Input AC **voltage** levels can be converted to digital numbers if the AC is *rectified* by a diode bridge circuit and then *filtered* by an RLC circuit. The resultant analog DC voltage can be converted to digital by an ADC. Figure 7.18 demonstrates the effects of rectification and filtering. Frequency information is lost.

Chips are available to convert AC **frequency** to analog DC voltage levels, and vice versa. To control AC frequency, a computer can drive a *DC-to-frequency* chip via a DAC. To "read" an AC frequency, the computer can use an ADC to read the output of a **frequency-to-DC** chip.

Modems convert digital DC signals to and from AC. They are used in data communication, and are discussed in that section of this chapter.

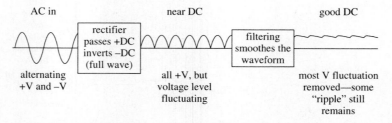

Fig. 7.18 Rectification and filtering of an AC signal

Serial I/O

The use of parallel I/O ports requires separate circuits for each bit of a data word. In serial I/O, *each bit of the data word is transmitted one at a time* on the same pair of conductors. Serial I/O sacrifices speed for a reduction in conductors. The speed sacrifice is often not noticeable to the user. Figure 7.19 shows how an 8 bit data word would be transmitted one bit at a time from a serial output port.

Serial ports are usually used for *inter-computer communications* including communication with several types of intelligent peripherals.

Where a controller is a significant distance from sensors and actuators (e.g., more than 2 meters), it is often cheaper and more dependable to use "intelligent" remote I/O modules than to wire each sensor or actuator back to the controller. A remote I/O module is directly attached to a group of local sensors and actuators. It contains sufficient computing power to exchange serial communication packets with the controller so that the controller can read sensors and write to actuators at the remote I/O module.

Configuration of programmable serial I/O ports involves more choices than in programming parallel I/O chips. The configuration of a serial chip is a large part of the **protocol** of communication that a computer will use.

Multiplexing

A computer can read several analog inputs via the same ADC, or receive serial inputs from several sources via the same serial port, if the input signals are **multiplexed.** Multiplexing inputs means switching the inputs to the input port one at a time, under computer control. Similarly, a computer can use a single DAC to output analog values to more than one output circuit, or a single serial port to output serial data to more than one output channel, if the outputs are multiplexed.

Analog values may have to be *held* via capacitors during multiplexing. Serial input must be either **"spooled"** into an external memory device or simply disallowed while the computer is switched away from a serial input channel.

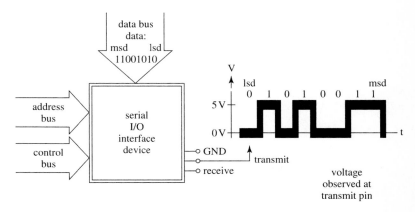

Fig. 7.19 Serial output

Communications and Protocol

Communication, as previously noted, is usually via serial I/O. The exchange of data may be asynchronous, or may be synchronous. In **asynchronous communications,** each data word is sent as a separate message. Asynchronous communication is often adequate if only two computers are connected via a set of conductors. Online programming and program monitoring is often done asynchronously with a single PC directly connected to a single PLC, a robot, or an NC controller.

Synchronous communication data messages consist of many data words that are preceded by a **header** containing information about the data **"packet,"** and followed by a **footer** containing error-checking information. Synchronous communication is appropriate where large amounts of data are to be transmitted quickly. Local area networks use synchronous data communications. [Although a recently proposed Asynchronous Communications Protocol (ATM) promises better data rates than synchronous protocols.]

The **RS 232 standard** is a widely accepted serial communication standard. The RS 232 standard dictates that a binary 1 should be sent as a negative 3 to 12 volts, and a binary 0 at plus 3 to 12 volts. The standard also specifies a 25 pin connector, even though only 3 pins are essential (ground, transmit, and receive). Some of the other pins have useful optional functions, such as for handshaking signals before the data transmission to ensure that the data path is free, or to verify that the data has been received in good order.

Figure 7.20 shows how two computers would have their serial ports connected via an RS 232. Other connections are necessary but have been omitted for simplification.

For serial communication at distances between 15 and 30 meters, the **RS 422 or RS 423 standards,** which call for better signal grounding, are better choices. The standard and options chosen become part of the communication protocol, and must be common at the data-sending and the data-receiving ends.

RS 485 is a standard with growing acceptance. It uses the improved grounding of the RS 422 standard, but allows the connection of multiple computers via the same set of four conductors. It can be used, therefore, in applications such as in in-plant local area networks. Most PLC local area networks use RS 485.

For long distance communication (over a mile), the above standards are not sufficient. One commonly-used solution is to convert an RS 232 output into AC signals that can then be carried by standard telephone lines.

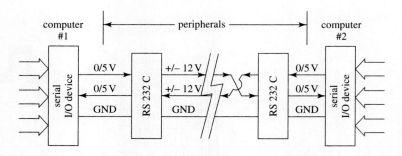

Fig. 7.20 Communication via an RS 232 interface

As shown in figure 7.21, a **modem** modulates (creates the AC) at one end, and a second modem at the other end demodulates the AC back into RS 232 (or equivalent) signals. Modern modems use a variety of frequencies, amplitudes, and phase shifting, with data compression techniques, to achieve transmission rates of over 10,000 bits per second (bps) over standard telephone lines.

Radio frequency modems, built into network adaptor circuit cards, can increase transmission rates into the 100 Mbps (millions of bits per second) range for in-plant communications. Network adaptors contain dedicated serial chips, and are connected to each other via specially shielded co-axial or "twisted-pair" cabling and connectors. The personal computer's serial port is freed up for other uses.

The communication hardware may be designed to use only two frequency pairs (one set for sending, another for receiving, with a communication controller to convert one pair of frequencies to the other). This is called a **baseband** communication network.

To allow more channels of communication on the same set of conductors, other frequency ranges may be available: one for standard telephone communication, a channel for video, another for a second communication channel, etc. This type of network is called a **broadband** network.

In simple frequency modulation (FM), only one bit can be sent at a time. One frequency means a binary one, the other frequency means a zero. High speed modems often combine FM with other modulation techniques, like amplitude modulation (AM) and/or phase modulation (PM). If, for example, the AC signal contains two frequencies and two amplitudes of AC, then there are four possible combinations of frequency and amplitude. Since it requires a two bit binary number to represent four combinations, the signal can carry two bits at a time! Each additional modulation capability doubles the number of bits that can be carried at one time. Modern computer networks always use bit-coding techniques of this sort.

A **carrierband** network does not use modems. It transmits serial binary data at voltages below 5 volts, at data transmission speeds into the 100 Mbps range. Small voltage levels can be switched faster than large levels, and switching rates are not limited by the period of an AC signal. The **integrated services digital network (ISDN)** is making slow headway toward acceptance as a standard for carrierband networking, but has powerful supporters in the telephone service companies.

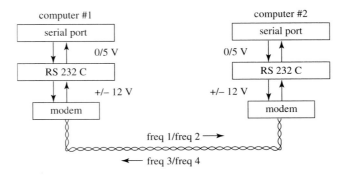

Fig. 7.21 Communication via modems

Serial binary data can also be coded into light. **Fiber optic** networks use light frequency and amplitude modulation techniques similar to those used in modems. The **Fiber Digital Data Integration (FDDI)** standard offers potential data transmission rates well into the hundreds of Mbps.

There are several **architecture** options for interconnection of cell controllers and central computers. Figure 7.22 demonstrates the hierarchical and three types of shared network.

In a **hierarchical** network, each link is dedicated to communication between only two computers (or "nodes"). Hierarchical architecture is successful where a clearly defined and non-varying communication hierarchy is apparent. Hierarchical networks are now rare.

Star network topology is similar to hierarchical, in that each link connects only two nodes, but differs in that some nodes act as message centers called "hubs," accepting and relaying messages to other nodes. Network hubs may not have any computing time left for other functions. There are many suppliers of star networks.

In **bus** networks, all nodes share the same set of conductors. Techniques are necessary to prevent "collisions" when two users try to send data simultaneously. One method of allowing shared access is the **Carrier Sense Multiple Access with Collision Detection (CSMA/CD)** bus standard. A computer that wishes to send first listens and if the line is free, transmits. While transmitting, the sender listens to ensure that no other computer has tried to send at the same time. If such a collision does occur, the transmission is immediately terminated and the sender waits for a short time before retrying the transmission. The popular Ethernet network uses CSMA/CD protocol.

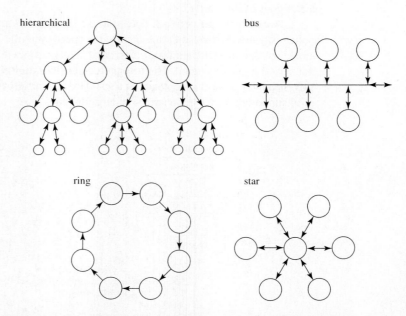

Fig. 7.22 Communication network architecture: hierarchical; bus; ring; star.

In **token passing** bus networks, each computer on the network gets its turn (token) in a pre-assigned rotation, and can transmit only at that time. GM's MAP network is perhaps the best known token bus network.

In **ring** networks, messages are passed from one computer to the next, in one direction, around the ring. The computer that is communicating passes control to the next computer in the ring when it has finished its communication. Computers waiting to originate communications must wait until they receive control. IBM is the major supporter of token ring networks.

Lack of standardization is the major problem in inter-computer communication (as the reader may have noticed by now). The large number of signal characteristics and topologies is just part of the problem. Manufacturers of processor chips, and the manufacturers of computer systems, have all developed their own instruction sets, memory access systems, and I/O access systems. A machine language program that works on one computer will not work on a different type of computer.

To get the same program to work on two different computer types, the high level language versions must be separately compiled into the machine language of each ... after the high level programming language supplier rewrites portions of the programming language program to run on the different type of hardware!

Even if the user wants the assorted computers in a factory to exchange only raw data, problems may exist. For example, one computer may use 32 bit data words and the other use 16 bits.

Computer equipment suppliers often offer **communication programs** that will allow one of their computers to communicate with another. These solutions do not usually even cover all the models within a manufacturer's line, let alone allow one brand of computer to communicate with another.

Communication software packages all require that both ends of the communication adopt a **common protocol.** Protocol specifies such factors as speed, data coding method, error checking, and handshaking requirements, and implies that the form of the data must be useful at both ends of the communication. There are no universally accepted choices among the protocol options, although there are several options commonly used.

Some standards are slowly evolving to allow computers to exchange data. A data "byte" is 8 bits. Text files are often encoded in 8 bit ASCII format. Asynchronous message formats have a manageable selection of options.

One of the efforts to promote standardization is that of the International Standards Organization (ISO) subgroup known as the **Open Systems International (OSI)** group. This group, with remarkable cooperation from industry, has developed the seven layer model for communications shown in figure 7.23, with specifications for the services to be provided by each layer. Although not all specifications have been agreed on, many of the most important have been adopted as standard. While not all suppliers have chosen to offer OSI-compatible controllers, the OSI effort has provided a pattern for computer compatibility. The U.S. government supports OSI compliance through its "**GOSIP**" requirements. Most suppliers now use the OSI pattern or something very similar.

The **OSI's seven layers** break up communication services into four **application service layers** and three **network service layers.** An "application" is taken as meaning any source of a request for network access at a computer. A computer operator or an

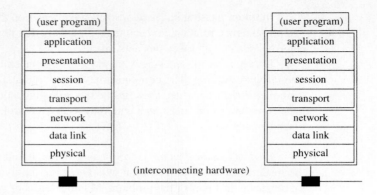

Fig. 7.23 OSI seven layer model

industrial control program would both be considered "applications." Commands from the "application" to the network's application services layer are recognized and responded to by the **applications** layer. The **presentation** layer translates data forms as required. The **session** layer at the sending computer works with the session layer at the receiving computer to transmit data only as fast as it can be received. The **transport** layer ensures that data is not lost or damaged in the lower layers. The four application service layers must exist at each node.

The three **network service layers** of programs may be provided by telecommunication companies, or may be built into a local area network. The **network** layer handles routing of messages via shared networks, and multiplexes messages so that a network can be shared by many "applications." The **data link** layer adds addresses to outgoing messages, examines all network data for messages addressed to this node, and does basic error checking. The **physical** layer inserts messages onto the shared network conductors and receives all message traffic, converting between binary data and whatever form data is in when being moved via the network.

There are situations where not all seven layers are required. Eliminating some reduces the hardware and software cost. If, for example, both computers can use the data in the same form, then the data translation function of the presentation layer is unnecessary. If the node's only function is to connect one type of network to another (e.g., token ring to CSMA/CD), then only the lower 2 or 3 layers are necessary.

General Motors' **Manufacturer's Automation Protocol (MAP)** is an attempt to hasten the fixing and acceptance of the OSI standard, and to force suppliers to adopt the standard. MAP's success has been somewhat limited, partly because GM has done what the ISO tried to avoid — antagonized large suppliers by adopting proprietary processes — and partly because the low volume of sales for MAP protocol conversion packages has kept the price higher than most users wish to pay. The CEO of one of the world's largest computer manufacturers, Digital Equipment Corporation, once asked why GM was telling DEC how to design computer networks. He added that DEC doesn't tell GM how to build cars! (DEC is a strong supporter of OSI.)

7.2.6 Programmable Logic Controllers (PLCs)

A programmable logic controller is a computer but operates somewhat differently than a personal computer.

The biggest difference between a PLC and a PC is that the PLC contains its operating system programs and application program in ROM memory. The user does not have to supply these programs on floppy or hard disk. In fact, the user can upgrade these programs only if the manufacturer makes new ROM chips available. A PLC "wakes up" when power is supplied as follows:

1. Like a PC, it retrieves a reset vector address from ROM, then starts to run the program that begins in ROM at that address. The first instructions in this first program have the PLC do a self-check. From here, the process is different from that of a PC.

The PLC does not have to load an operating system program, as it is already in ROM. The operating system does not have to load an application program, as it too is already in ROM. In fact, it would be difficult to differentiate between a PLC's "BIOS," "operating system," and "application" programming.

2. After the self-check, a PLC checks to see if there is an **EEPROM module** plugged into the CPU. If so, it reads the contents of the EEPROM into RAM. The EEPROM may contain a user program and/or data. Even if an EEPROM is not present, a user program may already be present in **battery-backed RAM**.

3. Some PLCs then run a user-written **initialization program,** even if the switch on the CPU is in the STOP position.

4. When the CPU is switched into **RUN** mode, most PLCs perform some form of **initialization procedure.** Some search for and run special user-written programs. Others set an internal memory bit for a short time, so that user-written programs can include initialization statements that are performed if that bit is set.

5. While in RUN mode, the PLC repeatedly executes a **scan sequence** that is much different from the way a PC would run a user program. The scan sequence, as shown in figure 7.24, next page, includes three steps, which are typically repeated (depending on the CPU and the program length) every 5 to 50 milliseconds. The steps are:

(a) *Input data from input modules into memory.* Current sensor values are read, and the data is put into RAM memory, often at memory addresses that correspond to where the input module is plugged into the rack it shares with the CPU. This section of RAM is called the **input image table**.

(b) *Run the user program.* A PLC program consists of a series of conditional statements. Most of these statements cause changes in the PLC's outputs dependent on the data in the input image table. A typical ladder logic instruction might look like this:

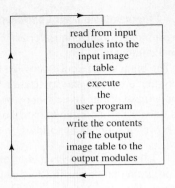

Fig. 7.24 A PLC scan sequence

which means: "if the switch connected at input module #4, terminal #3, is closed, actuate the solenoid valve attached at output module #6, terminal #2."

As the user program runs, it changes data in the part of RAM known as the **output image table.** PLC programming languages do not include "WAITFOR" types of statements, and most do not allow looping backward in the program, so the whole program can be executed in milliseconds.

(c) *Output data from the output image table memory to the output modules.* This step, executed after every complete execution of the user program, changes the actuator states.

6. When the PLC switch changes from RUN to **STOP,** the PLC stops repeating the scan sequence, outputs zeroes to all output modules, and *clears all "non-retentive" memory* locations. PLC programming languages include commands that allow the user to change data bits and data words in RAM, and commands that perform counting and timing functions. Some addresses for data words and counter and timer values are designated as **"retentive,"** and are not cleared by changing the RUN/STOP switch. They are maintained in battery-backed RAM when power is removed from the PLC.

Another difference between a PLC and a PC is that I/O devices for PLCs, called **I/O modules,** plug into a rack alongside the CPU module or into the side of the CPU, and are available only from the PLC manufacturer. Screw terminals are provided to connect sensors, actuators, or communication wiring. While typically much more expensive than the equivalent interface card for a PC, they are often cheaper in the long term because the user does not have to learn a new set of interface protocols for each module.

There are I/O modules for every type of sensor and actuator. Most I/O modules can be connected to more than one of the same type of sensor or actuator. I/O modules are available to handle communications between PLCs, other computers, and LANs. It is a good idea to check what types of I/O modules are available before selecting a make and model of PLC.

7.3 OTHER CONTROL COMPONENTS

This section will be devoted to **silicon-based devices.** Much of this text assumes at least a passing understanding of their properties. Readers with a good understanding of silicon-based devices can skip this section. Resistors, capacitors, and inductors are non-silicon devices. It is assumed that the reader already understands these devices.

Many silicon devices (diodes, transistors, etc.) can be used as digital or as analog devices.

Diodes consist of two layers of differently doped silicon: an N-doped layer and a P-doped layer. If a circuit provides more positive charge at the P end of a diode than at the N end, the diode is said to be forward biased. Figure 7.25 demonstrates that as forward biasing voltage increases from zero, current across the diode is nil until, at over 0.7 volts, it rises dramatically. The current rise with increasing voltage is so steep that resistance obviously has dropped to near zero. One major use of diodes is to rectify AC to DC.

Reverse biasing a diode yields a tiny current up to the breakdown voltage, beyond which the resistance to current is even less than in forward biasing. Reverse biasing a standard diode to the breakdown voltage will cause the diode to breakdown and burn up.

Zener diodes are intended to be used reverse biased. They provide voltage regulation by dumping charge at voltages over the breakdown voltage. The circuit on figure 7.26 will ensure that the load never sees more than the 6 volts that is this Zener's breakdown voltage.

The standard **bipolar transistor** revolutionized electronics.

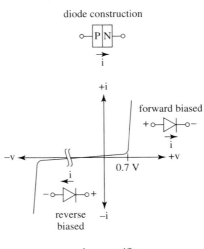

Fig. 7.25 Voltage-current relationship in a diode, and rectification of AC

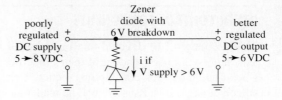

Fig. 7.26 Voltage regulation with a Zener diode

Used as a digital device, an NPN bipolar transistor is a switch that allows a large current to pass from the collector contact to the emitter contact when a small positive charge is supplied at the base contact. A PNP transistor allows current to pass from the emitter to the collector if the base is connected to a negative charge. A computer parallel port could use a transistor to switch a 12 volt motor on and off.

Used as an analog device, an NPN transistor's collector to emitter current increases proportionally with the base charge (see figure 7.27). Transistors thus can be used as analog amplifiers.

Field Effect Transistors (FET) are a variation on the standard bipolar transistor. The connection terminology is also different, as shown in figure 7.28. In an FET, charge at the Gate connection *reduces* current flow between Source and Drain connections. A bipolar transistor is thus normally open, while a field effect transistor is normally closed.

A single transistor can be used as an amplifier, but several transistors in series make a better quality amplifier. An integrated circuit containing many transistors and designed specifically as an amplifier is known as an **operational amplifier, or "op amp."** Op amps are designated by the symbol shown in figure 7.29 (a). There are two power supply contacts, one for positive power, the other for negative power supply. These contacts are

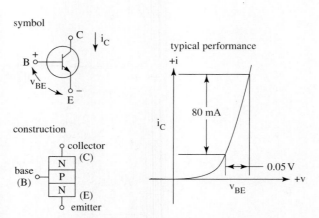

Fig. 7.27 NPN transistor base-emitter voltage versus collector-emitter current relationship

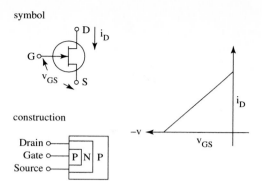

Fig. 7.28 N-channel field effect transistors

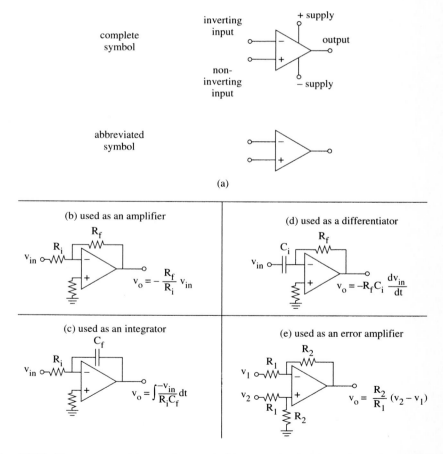

Fig. 7.29 The op amp: (a) complete and abbreviated schematic; (b) as an amplifier; (c) integrator; (d) derivative amp; (e) error amplifier

7.3 Other Control Components **223**

often not shown on a schematic drawing. There are **two input contacts.** Output of the op amp is proportional to the difference between the voltages at these two inputs. If, as is frequent, the non-inverting contact is connected to ground, then output is inversely proportional to the voltage at the inverting input.

One feature of the op amp is that it draws next to no current from the input circuit (it has almost infinite input impedance), while providing almost zero impedance to the output current. The op amp therefore, has little effect on the input circuit and does not have any undesirable effect on the output circuit, so it is an excellent amplifier.

An op amp can be used to **amplify** current or voltage. It provides gains in the thousands! The output's maximum and minimum values are limited by the supply voltages. An op amp is said to be **saturated** when its output is at its maximum or minimum. Proportionality between input and output is poor at large gain values, so gain is usually sacrificed for quality by using the op amp in a circuit with feedback. Figure 7.29 (b) shows how input resistors and a feedback resistor, between the inverting input and the output, control the amplification.

An op amp also can be used as an **integrator** (figure 7.29 (c)). The output of an integrator remains unchanged if there is no input voltage at the inverting input. If a voltage is applied at the inverting input, the output changes for as long as the input is applied, at a rate proportional to the input voltage.

Used as a **derivative amplifier** (figure 7.29 (d)), the op amp provides an output only while the input is changing. The output is proportional to the rate of change.

The **error amplifier** circuit (figure 7.29 (e)) provides an output proportional to the difference between the inputs. Used as an analog device, this circuit allows all the resistor values to be chosen the same so that output will be equal to the difference between the inputs. Used as a digital **comparator,** there is no feedback resistor. A comparator will always be saturated at either full positive or full negative output. If only one of the two power supplies is connected, the comparator's output will be either fully on or fully off.

A **thyristor** is a four layer device that starts to conduct under either of three conditions: If there is a forward biasing voltage across the anode (A) to cathode (K), a pulse of sufficiently high charge at the gate (G) will start it conducting. Conduction can also begin without a gate pulse, if the forward biasing voltage is very high. Conduction may begin if the anode to cathode voltage increases rapidly. Conduction continues until the device ceases to be forward biased.

A **silicon controlled rectifier (SCR)** is a thyristor that uses the gate turn-on method. SCRs are used in DC motor speed controllers to rectify AC and control motor speed. The circuit in figure 7.30 controls output DC level by controlling the delay between when the SCR is forward biased by the AC, and when the gate potential reaches a sufficient value to start conduction. Increasing the resistor value decreases the rate at which the positive AC supply can charge the capacitor, delaying the time when the voltage at the SCR gate reaches "triggering" level, and thus reducing the amount of current allowed to pass during this AC cycle. See chapter 3 for more on SCR motor control.

Gate turn off (GTO) thyristors exist. They have a fourth contact, which is used to stop conduction even if the thyristor remains forward biased. GTOs are not in common use.

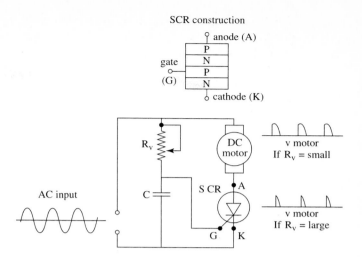

Fig. 7.30 SCR delay control circuit

Another thyristor variation is the **triac,** which is turned on like a standard thyristor (gate pulse, strong bias, or rapidly changing bias), but can be caused to conduct in either direction. Conduction stops when the forward or reverse biasing voltage drops to zero.

7.4 ANALOG CONTROLLERS

There was a time when analog controllers, also called **analog computers,** were considered the best choice for control applications, particularly for process control. Chapter 6, devoted to servosystems, demonstrates that analog terminology and block diagrams still predominate in the area of process control.

Analog computers are not reprogrammable. This means that an analog computer for a given function must be **custom-designed** and built for that function. Some limited reprogrammability can be designed in if the circuitry includes variable resistors, variable capacitors, or variable inductors. Transistor radios, for example, are analog computers with variable capacitors that allow the radio to be "re-programmed" to receive different stations. The radio can't be "adjusted" to be used as a calculator. Significant reprogramming of an analog computer would require that it be rebuilt.

With the ever-dropping price and increasing speed of digital computers and interface devices, analog computers are becoming less common.

There are still applications where analog computers are desirable, such as where **speed and resolution** of the control are critical.

7.4.1 Analog Controllers Under Digital Control

Analog controllers are unlikely to disappear in the foreseeable future. Their speed and precision make them ideal for certain tasks, where they will frequently be slaves of digital computers.

Robot controllers, for example, include analog servo-controllers for DC motors, built into the same case as the digital computer. The analog controllers are monitored and controlled by the digital computer.

Analog controllers are shrinking in size. They are now available in integrated circuit form, with connections for user-selected resistors, capacitors, and inductors. Some have data, address, and control bus connections for use as I/O devices in a digital computer.

7.5 SUMMARY

For Computer Integrated Manufacturing, the user must **interface a central controller to peripheral equipment** such as other computers, robots, numerical control equipment, programmable controllers, cell controllers, or to groups of sensors and actuators.

The **host controller** is often a mainframe computer. It provides memory and software to store and manipulate the manufacturing database, database management programs, scheduling programs, and network control programs.

Automated manufacturing requires digital computers. Digital computers contain a **central processing unit (CPU), memory** for programs and the data it must process, and **input/output ports** via which they receive and transmit data. In this chapter, we examined the architecture of a simple microprocessor (MPU), a CPU on a chip. We also examined how the MPU used **several layers of programs** to run high level programs. We examined the slightly different way in which a PLC's memory is organized and how it runs its user program.

Parallel I/O is sufficient for controllers that only monitor digital sensors and control digital actuators. Signal isolation and line driver chips are often required. **Analog I/O** requires the use of ADCs and DACs, but pulse width modulation, DC to frequency, and conversion frequency to DC are also available. **Multiplexing** of I/O lines can reduce the number of analog I/O chips required.

Serial I/O is preferred for inter-computer **communications,** as it requires fewer conductors. Serial communications can be asynchronous, but synchronous communication is faster. Serial communications usually require the addition of extra hardware such as **RS 232 drivers, modems, and/or local area networks.** There are several proprietary LANs available, with star, bus, or ring architecture, some with "tokens" used for network access and some using CSMA/CD protocol for network access. The standards being developed by **the OSI subgroup of the International Standards Organization** are helping to force suppliers to move toward interconnectivity.

Several silicon-based devices were discussed. **Diodes, transistors, FETs, op amps,** and **thyristors** such as the SCR are used extensively as digital and as analog components in automation. The **op amp** is the heart of the analog computer, which is now usually seen as a slave controller to a digital computer. Simple versions of analog computers are used as interface devices between digital computers and peripherals.

REVIEW QUESTIONS

1. What items must be added to a standard personal computer before it can be used as an industrial controller? Which peripherals may not be necessary?
2. What is an industrial cell controller used for?
3. What type of data might be stored in the database in a manufacturing plant's central computer?
4. What are the 3 buses of a computer, and what sort of information does each carry?
5. What are the three types of computer components connected by buses, and what is each for?
6. Which type of computer memory is lost when the power goes off? Why do many industrial controllers such as PLCs not lose their user programs from this type of memory when turned off?
7. What is the difference between EPROM and EEPROM memory? Which is commonly used to store user programs in PLCs?
8. Which of a microprocessor's registers stores numbers to be manipulated?
9. What is the purpose of a microprocessor's program counter register?
10. List, in the order that they are run, the levels of software required to run a user program written in (interpreted) BASIC on a personal computer.
11. Which is faster, serial or parallel data communication? Why? Which is more frequently used for intercomputer communications? Why?
12. Identify the signal conversion processes required in the following cases:
    a) A local controller (digital computer) is reading a low power digital switch.
    b) A local controller is reading a value from an analog sensor.
    c) A local controller controls the on/off state of a DC motor.
    d) A local controller is sending data to another local controller via a baseband local area network.
13. Describe how a signal in one circuit can control another circuit's on/off state via an optical isolator. Why is the optical isolator necessary?
14. Why might an op amp be used in a sensor's circuit?
15. Define "discrete digital."
16. How many steps would it take for the following analog to digital converters to complete a conversion? If you can't answer, describe why not, clearly identifying the process involved that makes the answer impossible.
    a) an 8 bit successive approximation ADC
    b) an 8 bit flash ADC
    c) an 8 bit dual ramp ADC
17. If you have only one ADC, but must read analog values from several sensors, what additional equipment do you need?
18. What does an RS 232 interface do?

19. What does a modem do?
20. Use voltage versus time graphs to show what the number 10010100 would look like as it is sent:
    a) from a serial I/O device to an RS 232
    b) from an RS 232 to a modem
    c) from an FM modem to another FM modem
    d) from a modem to an RS 232
    e) from an RS 232 to a serial I/O device
21. If you wanted to double the speed that information can be sent over the lines connecting two RF modems, without increasing the frequency at which the modems changed their output signals, what would you look for in new modems?
22. Which can carry more data, a baseband or a broadband data network? Why?
23. ISDN networks are neither baseband nor broadband. What kind are they? Describe the serial signals on this type of network.
24. Describe two different sharing methods that would allow several computers to share a bus architecture network. Which of these is the method specified for a MAP network?
25. Which of OSI's seven layers adds and recognizes network addresses?
26. How does a standard diode differ from a Zener diode?
27. How would a transistor be used differently in a digital computer from how it would be used in an analog computer?
28. If R_i were 1000 ohms, what would R_f have to be in an op amp amplifier circuit to achieve a gain of 2?
29. Using graphs of voltage versus time, show the input and output of an op amp integrator if, at time zero, both input and output are zero, but at time 2 seconds, the inverting input became 2 volts, and at time 5 seconds it went to −1 volt for 4 seconds. ($R_i * C_f = 1$, and saturation is not exceeded.)
30. In what sort of application is an analog controller still superior to a digital controller?

PART D

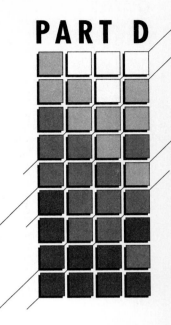

AUTOMATED SYSTEMS

8	ROBOTS	230
9	THE AUTOMATED WORK CELL AND ITS FUTURE	255
10	AUTOMATION: WHEN? HOW? AND OTHER NON-TECHNICAL ISSUES	273

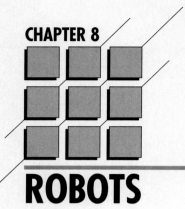

CHAPTER 8

ROBOTS

CHAPTER OBJECTIVES

After completing this chapter, the reader will be able to:

- Discuss the differences between servo-robots and pick and place robots.
- Distinguish among cartesian, cylindrical, spherical, articulated, and SCARA robot geometry.
- Describe roll, pitch, yaw, and Z axis degrees of freedom.
- Discuss the relative merits of electrical, pneumatic, and hydraulic power supplies for robots.
- Describe a variety of optional end effector choices.
- Identify the components of a robot controller, and the function of each component:
  - the power supply
  - the processor
  - the memory, including ROM routines for:
    - servocontrol
    - motion control
    - program entry and execution
  - I/O ports for:
    - axis control
    - parallel and analog peripheral control
    - offline programming and other communications
- Define the level of sophistication of a robot in terms of controller strengths for:
  - path control programming
  - programming language capabilities

8.0 ROBOTS

Robots are usually thought of as actuators, in that they do work. These "actuators" can perform marvelously intricate tasks in response to a single computerized request. Robots are often included in Flexible Manufacturing Systems (FMS) and in Computer Integrated Manufacturing (CIM) systems.

Anyone who has been faced with the task of purchasing or installing a robot soon finds that there is more to a robotic actuator than there is to most other actuators. This "actuator" consists not just of electric or fluid motors, but includes built-in sensors, computers, software, and a bewildering selection of options and interfaces for other equipment. In this chapter, we will attempt to make sense of the choices the designer and user are faced with.

As shown in figure 8.1, a robot consists of:

- the actuator, called the **"arm"** in this chapter to distinguish between it and the individual actuators that move it
- a **power supply,** often built into the controller box as in the photograph

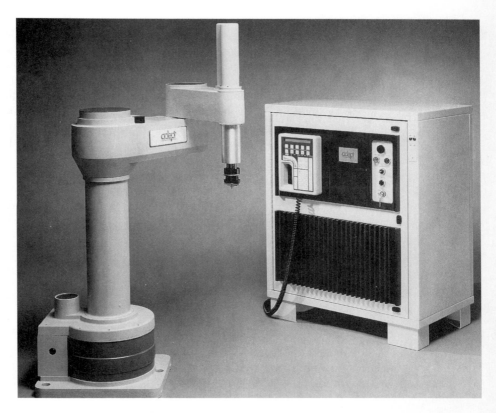

Fig. 8.1 Components of a robot. (Photograph by permission, Adept Technology, Inc., San Jose, California.)

- a **controller,** which includes a pendant for operator control. The pendant is hung on the front of the controller in the photograph

We will be examining each of these components in more detail in later sections.

8.1 ROBOTS AND NEAR-ROBOTS

The term *robot* originated in a Czechoslovakian play about humanoid machines used as slaves. (The machines eventually rebelled.) "Robot" now means a reprogrammable machine with several independent degrees of freedom, which can exchange information with other devices. This definition also implies that the computer that controls the degrees of freedom is a part of the robot. A true robot can be programmed to move to any of the locations between the limits on each of its degrees of freedom. After this section, the rest of this chapter will deal with modern servocontrolled robots, which fulfill the requirements of this definition.

Some early robots, still widely used in industry, did not fully meet the above definition, but are still called robots. Figure 8.2 shows a robot of this type. These near-robots are now often called **pick and place** robots, and can be a better choice than true robots in some situations.

Pick and place robot arms consist of at least two degrees of freedom and some type of end-effector (often a gripper custom-built for a particular application). Each degree of freedom is a pneumatic actuator built into a precision linear or rotary slide mechanism. Each pneumatic actuator is controlled by a simple solenoid valve (not included with the robot), so that it can be driven in one direction or the other direction. A pick and place robotic actuator cannot stop between the endpoints of its range. These robots are sometimes called "bang-bang" robots because they can be moved only to positions "bang" against their stops. These robots are often used to repeatedly move parts from one fixture to another, hence the name "pick and place." Sensors (to detect when an actuator has been driven to a stop) are rarely included. Users must install their own sensors.

Pick and place robots are not usually supplied with controllers! The most commonly-used add-on controllers for pick and place robots are PLCs (programmable logic controllers).

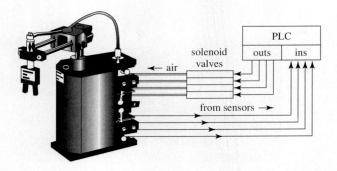

Fig. 8.2 A pick and place robot. (Robot by permission, Jergens Inc., Special Products Group, Cleveland, Ohio.)

A PLC can actuate the valves of a robot in the sequence required. The same PLC can be programmed to examine sensors (if present) to ensure that each actuator has driven to its limit before the next is actuated.

Because pick and place robotic arms are not complex, they can be built inexpensively and with better repeatability than most servo-driven robots.

Pick and place robots can be purchased as pre-built units, or as individual linear or rotary modules that can be assembled for the specific combination of degrees of freedom required.

8.2 ROBOT GEOMETRIES

Robots have more than one degree of freedom. A **degree of freedom** is the ability to move about a linear or rotary axis. Each axis of a true robot has a separately controllable actuator and a sensor, and circuitry and programming to control that axis. Robot speed is best quoted in terms of the linear or angular maximum speed attainable per axis, but is sometimes quoted as the maximum attainable velocity at the robot gripper when all axes are moving together.

8.2.1 The Three Major Axes

Robots are classified according to the particular mixture of their three primary axes. In figure 8.3, we see that robotic arms may be either:

- **cartesian** (three linear axes)
- **cylindrical** (two linear and one rotational axis)
- **spherical** (one linear and two rotational axes)
- **articulated** (three rotary axes)

The most successful of these three robot geometries has been the articulated type, perhaps because we humans (with our rotary joints) expect robots to perform motions suited to our limbs. Robots with linear axes have also enjoyed some success because humans tend to define space in terms of three linear axes, and can easily design robotic workcells on the drawing-board in this linear axes system.

The earliest servocontrolled robots were of the spherical type. These and the cylindrical type robots are still popular because these robots allow robot manufacturers to use well-understood rotary actuators for the major axes.

Perhaps the most successful robot geometry has been a variation on the articulated geometry, the **SCARA,** or selective compliance articulated robot actuator, which has three **vertical** axes of rotation. (The standard articulated geometry has only one vertical axis of rotation.) The SCARA robot has been so successful because we tend to lay out workcells on a flat surface (the drawing-board influence again), and because the SCARA's vertical axes of rotation are ideally suited to allow (have compliance for) minor inaccuracies in this two-dimensional world.

We saw in chapter 6 that proportional servosystems use little force to try to correct small errors. This means two things. It means that if a SCARA robot tries, as in

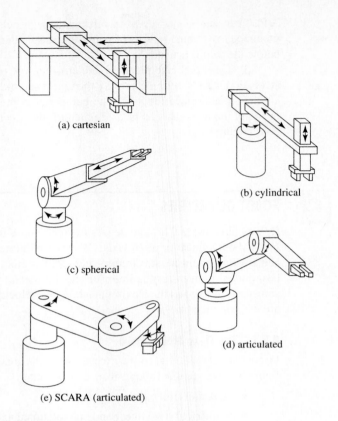

(a) cartesian

(b) cylindrical

(c) spherical

(d) articulated

(e) SCARA (articulated)

Fig. 8.3 Robot geometry types

figure 8.4, to insert a chamfered shaft into a hole that is slightly out of place, the shaft can pull the robot into alignment with the hole. Little resistance will be offered by the servo-controller. Non-SCARA robots attempting the same insertion would twist about their axes of rotation in undesirable ways, as shown in figure 8.4.

Selective compliance also means that error in the robot servocontrolled positioning is more easily tolerated. SCARA robots moving components to vertical holes cannot accidentally rotate the components away from the vertical condition. Positional error in a non-SCARA articulated robot will inherently mean that the same component would be shifted from vertical.

There has been much work done to combine axes of rotation, similar to the way that the human backbone and muscle system allow us to rotate about more than one axis. Complex universal joints have been installed in some robots to solve this problem. Some robots, for example, have an arrangement of cables that, when selectively drawn tight, bend a series of joints in much the same way that the human spine bends. Despite these efforts, there have been no popular robots in which the three primary axes geometry has differed from the four types discussed.

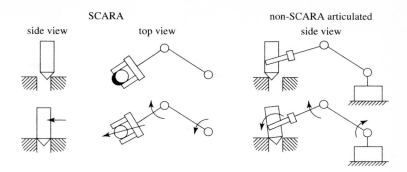

Fig. 8.4 Selective compliance

8.2.2 Additional Degrees of Freedom

All robots require an end effector of some sort to be useful. For maximum robot versatility, additional degrees of freedom are needed at the end effector, just as humans require several degrees of freedom in our wrists. Each added degree of freedom is also a source of positioning error, not to mention increased cost in manufacturing and maintenance, so a balance must be struck between too many and too few degrees of freedom.

Many robot suppliers have found that the six degrees of freedom robot, shown in figure 8.5, provides a good balance. This robot has its three primary degrees of freedom arranged to yield an articulated geometry. The additional three degrees are:

- wrist **roll** (a rotation about the axis through the gripper)
- wrist **pitch**
- wrist **yaw**

The pitch and yaw axes are axes of rotation perpendicular to the gripper axis, and perpendicular to each other. Not all articulated robots have six degrees of freedom. The yaw axis is often considered unnecessary.

These three extra degrees of freedom are rarely added to a SCARA robot. Non-vertical axes of rotation would destroy the "selective compliance." SCARA robots do, however, require degrees of freedom beyond their three primary rotational axes. All SCARA robots are provided with a fourth, linear degree of freedom.

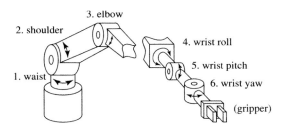

Fig. 8.5 Six degrees of freedom

8.2 Robot Geometries **235**

As can be seen in figure 8.6, the additional **linear** degree of freedom, called a Z-axis, is also vertical. It allows the robot to lift and lower the flange to which its gripper is attached or, more rarely, to lift and lower the arm itself.

Note that the degrees of freedom discussed so far end at the **gripper flange.** This is where the user must attach an "end effector" (often a "gripper").

End effectors, discussed in section 8.4, are usually designed and built by the end-user of the robot, although many robot suppliers can provide basic grippers.

The end effector is usually not included in the count of a robot's degrees of freedom. The exception to this rule is if the gripper is a **servogripper,** provided by the robot manufacturer, and controlled by circuitry included in the robot controller; it can be and often is listed as an additional degree of freedom. The purchaser should be aware that a six degree of freedom robot, if supplied with a servogripper, may lack wrist yaw.

 ROBOT POWER SOURCES

To be able to accurately move to specified locations, a robotic arm needs actuators in each of its axes that can be precisely controlled. Actuators that can be servocontrolled are required. The two previous chapters, dealing with servocontrol and with controllers, discussed how digital computers can be used to control analog processes such as positioning of a robotic actuator. It has only been since the arrival of digital electronic controllers that robots have been practical.

8.3.1 Electric Motors

Because modern robots were made practical by those who understand electricity and electronics, it is only natural that the vast majority of robots on the market today are powered by **electric motors.** These robots are reasonably fast, precise, and are cleaner and quieter in operation than most non-electric robots. By quiet, we mean low in audible noise. Designers of electronic controls can correctly point out that electric motors are a source of electromagnetic noise. Nonetheless, robots with electrically powered actuators are at the forefront of modern robotic technology.

Some early electrically powered robots where built using **stepping motors,** and were driven by open loop controllers. In open loop control (see figure 8.7), the robot controller has no sensor to verify that the actuator has moved its axis to the desired position.

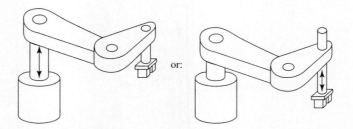

Fig. 8.6 The SCARA Z-axis

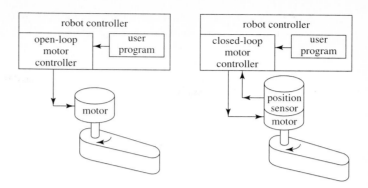

Fig. 8.7 Open versus closed loop control of electric robotic actuators

Since a stepper motor is supposed to move a fixed angle for each electrical pulse it receives, it should not be necessary to use a sensor to measure the axis's position if the pulses given the motor are counted. In fact, there are several reasons why a pulse may not result in a move (e.g., arm jammed or inertia exceeds motor torque).

Stepper motor driven robots have all but disappeared, but stepper motors have improved over the years and may show up in robots again, with closed loop control next time.

Most modern electrical robots have closed loop control systems (with feedback from built-in sensors at each axis) driving **DC permanent magnet** or **brushless DC motors.**

DC motors generate a torque proportional to the DC current they receive. The major drawback of traditional DC motor design is that the motor requires a light armature to be able to accelerate quickly, whereas high acceleration torque can be generated only if there are heavy conductors in the armature to allow high current rates! Even today, most robots incorporate fast motors with light armatures, which are highly geared-down to optimize the torque/acceleration balance.

Brushless DC motors are an improvement over standard DC motors. These motors are built somewhat differently from traditional DC permanent magnet motors. In chapter 3 we saw that the armature of a brushless DC motor is stationary, and the field, with its permanent magnets, rotates. Brushless DC motors can have heavy windings in the armatures (since the armatures do not have to be accelerated), so they can develop high torques. These motors are used in a new generation of **direct drive** robots and eliminate gearing. They also eliminate the troublesome commutator required to supply DC to a rotating armature. **AC induction motors,** too, have been used in robots, but do not seem to have been as successful as DC motors, yet.

8.3.2 Pneumatic Actuators

Many manufacturing plants use compressed air as an inexpensive power source. It is only natural, therefore, that many of the earliest robots were powered by **pneumatic actuators** (see figure 8.8), which used the available compressed air as a power supply. A **filter-regulator-lubricator (FRL)** is required between the air-supply line and the **valves,** which are required to control the robot.

The biggest disadvantage with compressed air as a power source is that it *is* compressible. Air power is not able to accelerate (or decelerate) heavy loads as quickly as non-pneumatic actuators can. Attempting to accurately control the position of an actuator by metering air volume to the actuator is also frustrating.

In a closed loop control system (where air is supplied to the actuator until it reaches its desired position), the inertia of the actuator is sufficient to lead to unsatisfactory overshoot.

Pneumatic actuated equipment is often noisy, due to the frequent need to exhaust high pressure air. Pneumatic actuators can be too dirty for some applications, because the air that is exhausted to the workplace usually carries a fine mist of oil intended to lubricate the pneumatic actuators. If exhaust air is contained, the noise and oil problems can be overcome. Another irritating feature of compressed air is that it gives off heat when it is compressed, and absorbs heat when exhausted.

Servocontrolled pneumatic robots have thus all but disappeared. (They may reappear, however. Some very good servocontrolled pneumatic actuators have recently been made available.) **Pick and place robots,** on the other hand, are pneumatically actuated and are still very popular. Many non-pneumatically powered robots include air-powered grippers.

8.3.3 Hydraulic Actuators

Figure 8.9 shows the components in a hydraulically powered robot. In addition to the arm and the computerized controller, the robot must also include a pump, a reservoir, a filter and pressure regulator, and servovalves in the power supply section.

Hydraulically actuated robots have enjoyed only limited success, despite some real advantages.

Unlike air, oil volume does not significantly change with pressure. It can be pumped at high pressures (3000 to 5000 psi is not unusual), and therefore can be and is used to power very strong and very fast robots (compared with electrically-actuated or pneumatic robots).

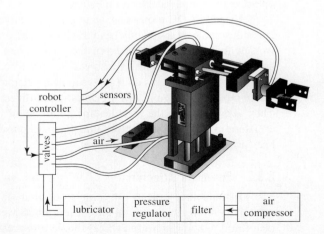

Fig. 8.8 Pneumatic robot with power supply and auxiliary components. (Robot by permission, Barrington Automation, Crystal Lake, Illinois.)

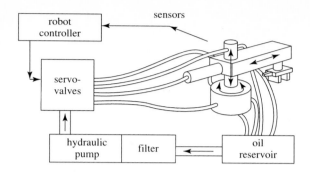

Fig. 8.9 Hydraulically powered robots

Driven by high oil pressures, hydraulic robots can correct small positional errors quickly and precisely. Oil is metered to the various actuators by very precise electrically controlled servovalves, which open proportionally to the electrical signal they receive.

Hydraulically-actuated robots are quiet, relative to pneumatic robots, since the only noise source is the motor powering the hydraulic pump (which can be enclosed in a sound absorbing box).

The drawbacks of hydraulic power are that:

- unlike air and electricity, pressurized oil is not available in most manufacturing plants, so the cost of an expensive hydraulic pump must be justified
- the pump is noisy, and occupies valuable floor space
- maintenance of hydraulic equipment can be messy, if highly skilled maintenance workers are not employed: When a hydraulic hose is disconnected, oil leaks out; if the connection is not tightly sealed during operation, oil will continue to leak
- if a hydraulic hose or connector should break while the system is pressurized, 5000 psi oil can be dangerous
- because hydraulic equipment is not as popular as electrical or pneumatic, hydraulic components are more expensive, so the hydraulic robots are expensive

A final reason why hydraulic robots are not as common as electrical robots, despite their obvious speed and power advantages, is that their designers must know hydraulic theory as well as electronic theory. The reader can easily verify that hydraulic robots often lack the more recently developed features that electric robots seem to incorporate as soon as they are developed.

8.4 COMPONENTS OF A COMPLETE ROBOT

The introduction to this chapter said that a robot consists of an **arm,** a **power supply,** and a **controller.** These three components are not very useful in themselves. To be useful, the robot, as shown in figure 8.10, must also

- have an **end effector** (usually user-built), and
- be built into a system, often containing special **peripherals,** which the robot controller can control or at least exchange information with. The workcell shown in figure 8.10 is explained in detail in chapter 9, The Automated Workcell and Its Future.

We will now add end effectors and peripherals to our list of components of a complete robot, and examine all of these components in more detail.

8.4.1 The Arm

In discussing robot geometry and power sources, we have covered a good deal of what can be said about the arm. Suffice to say that robot arms include an assortment of rotary and/or translational axes, driven by electric or fluidic actuators, and with a built-in position sensor at each axis.

The sensor is essential to the servocontrol of each axis. The accuracy of the axis is limited by the resolution and repeatability of the position sensor. Most robots use incremental optical encoders built into each joint (see figure 8.11) to sense the axis's position. Rotary resolvers and synchros are used in some other robots.

It is important to note that sensor placement (at the actuator), is less than ideal. The user of a robot wants the robot's payload to be positioned accurately, yet the actual payload position is not directly measured by the sensor. Variations in payload weight, orientation, or inertia during acceleration will lead to variations in the elastic deformation of the arm, which cannot be detected by sensors placed at the other end of the elastic link.

Robotic arms typically come with base flanges so that they can be bolted down to a worktable, and with gripper flanges to which end effectors can be mounted.

Most robots come complete with some built-in air lines, sometimes with built-in air valves, and occasionally with built-in wiring to operate the end effector. Sometimes the built-in valves and electric circuits are hard-wired right back to the controller and respond to special commands in the robot programming language.

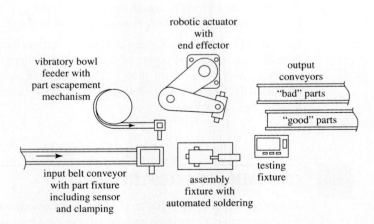

Fig. 8.10 A complete robotic workcell

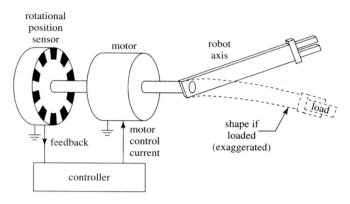

Fig. 8.11 Robot axis position sensor, showing undetectable elastic deformation

8.4.2 End Effectors

The end effector is usually designed and built by the end-user. Robot manufacturers and some manufacturers that specialize in end effectors can supply basic end effectors, but except for welding, spray painting, and a few such applications, it is usually better for the user, who understands the application best, to design his or her own.

The **gripper** is the most common form of end effector. Grippers, as shown in figure 8.12, often have two fingers, but may have more. They can be designed so that each finger hinges on a single pivot (called **angular** or "radial" or "birds-beak" type), or designed so that fingers remain **parallel** to each other as they open and close.

Grippers can be designed such that the "fingertips" can be changed to match the shape of the workpiece.

More often than not, grippers are pneumatically operated, and their closing force can be controlled only by controlling air pressure. Some grippers are actuated by small electric or hydraulic servo-actuators so control of the gripper's closed position and closing force is possible. Sensors on the fingertips may then be useful.

To pick up flat materials, **vacuum cups** are often incorporated into end effectors. Metal parts may be lifted with **electromagnetic** end effectors. Vacuum and electromagnetic end effectors can be quick to actuate. These and other end effectors are shown in figure 8.13.

The term *end effector* rather than *gripper* implies that the robot doesn't necessarily have to grab the workpiece to do work on it. A variety of end effectors allow robots to **spray paint, weld, cut, install nuts and bolts, etc.**

Dual purpose end effectors, like those shown in figure 8.13, with independent control signals, can cut down a robot's cycle time by allowing the robot to pick up two dissimilar parts from separate supply locations and deliver them together to a final destination.

End of arm fixturing that allows the robot to pick up selected end effectors from a rack can make for a truly flexible manufacturing cell, as long as the fixturing is not so heavy that it reduces the payload capability of the robot too much, and providing frequent end effector changes does not increase cycle time too much.

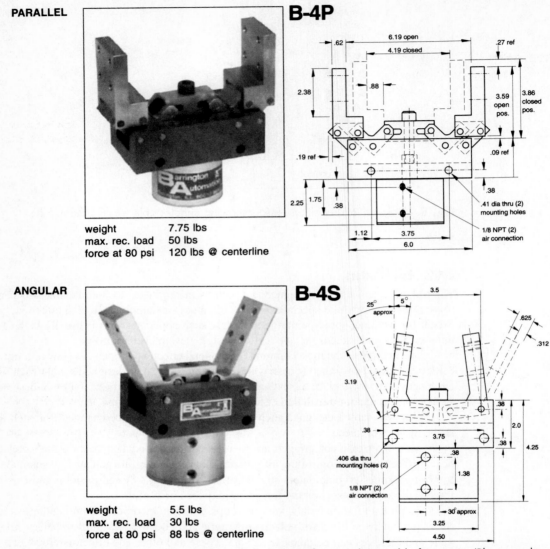

Fig. 8.12 Basic two finger, angular, and parallel grippers, for interchangeable fingertips. (Photographs and diagrams by permission, Barrington Automation, Crystal Lake, Illinois.)

8.4.3 Peripherals

One of the strongest arguments for using robots in an automated process is that a robot can easily be programmed to respond to changing conditions. To take advantage of this feature, the designer of a workcell containing a robot must design peripherals to be controlled or monitored by the robot. Other chapters in this book describe sensors, actuators, and process controllers that can be interfaced to the robot controller. These peripherals "become" part of the robotic system.

In a CIM system, the robot itself is a peripheral of the host computer! Interconnection of the robot controller to its peripherals, and to other controllers, was discussed in the previous chapter about controllers. It will be discussed again in this chapter, when we examine robot controllers, and will be discussed a final time in the next chapter when we examine complete automated work systems.

8.4.4 The Power Supply

Power supplies condition the available power (usually AC power purchased from the local electric power utility) so that it can be used in the robotic workcell.

At least part of the power supply supplied with the robot may be built into the same case as the controller. This is the portion of the power supply that conditions the **electric power.**

The power supply must **filter** undesirable spikes out of the input power, and perhaps provide a battery **backup** in case of power interruption. The input AC power must be **rectified and regulated** to provide low voltage DC to run the computer, and often **regulated to several other DC levels** to provide power to motors in the arm and to sensors.

Hydraulic pumps are also considered power supplies, when they provide the power to move hydraulic robots. The hydraulic power supply unit requires pressure control

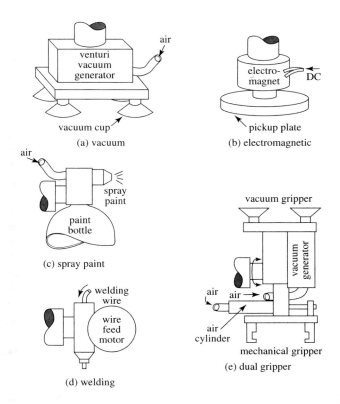

Fig. 8.13 Other miscellaneous end effectors

valves, filters, and oil reservoirs. These hydraulic power supply units are provided in modules remote from the electric power supplies, for obvious reasons.

Pneumatic robots require pressurized air to operate. The **air compressor** is not provided with the robot, and is not considered part of the robot.

Filters-regulators-lubricators (FRLs) should be considered an essential power supply component for a pneumatic robot. Filters catch dirt, pressure regulators limit the air pressure, and lubricators insert a fine mist of oil into the air stream. These components are usually not provided with the pneumatic robot, and not all new users of compressed air recognize their importance, so sometimes they are not installed when they should be.

Another important pneumatic power regulator is the adjustable **flow control valve**, used to limit the speed of operation of pneumatic actuators.

8.4.5 The Controller

The controller contains almost all the technological advances that have made robots viable. Figure 8.14 shows the **digital computer** that is at a robotic controller's core. The computer consists of:

- one or more **microprocessors** as a central processing unit
- **memory** containing programs written by the robot supplier, and with room for programs and data entered by the end user
- **input/output devices** through which the computer receives and transmits information

The average robot user must be at least familiar enough with the controller to "teach" the robot important locations, to store and run robot routines, and to connect peripherals.

Central Processing Unit

Robots became popular when 8 bit computers, such as the popular Apple 2E, were the norm. There are still robots available that have controllers built around an 8 bit

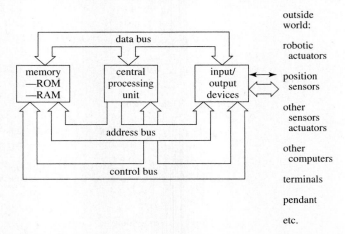

Fig. 8.14 The controller

microprocessor such as Motorola's 6800, or Zilog's Z-80. The disadvantage of these early robot controllers was that, besides being slower than more recent microprocessors, they had severe limitations on the amount of memory they could address. The more features the robot designer included in memory, the less memory was available for user-written programs and positions in the workspace that the robot had to remember (called "taught points").

Many robots today have 16 bit microprocessors, often with co-processors for more efficient calculation. A commonly-used microprocessor is Intel's 8086 or 8088 (which was the original processor in the highly-successful IBM PC family of computers), although other 16 bit processors are in use. The increased speed and increase in address-able memory allowed better control of arm velocity and accelerations, and allowed more features in the robot programming languages. The degree to which industry embraced the IBM PC family of computers created an industry standard, which meant that for the first time it was possible to expect computers to be able to exchange information with each other. Offline programming, where a program could be written at a desktop computer then run on the robot controller, became a common feature during this period.

Thirty-two bit microprocessors are at the heart of the latest generation of robot controllers, again because they offer increased speed and memory size. A 32 bit computer is better suited to the memory needs of continuous path robots (which must remember large numbers of locations), and is required for the added manufacturer-supplied, built-in features demanded by users. Unfortunately, there is no single favorite 32 bit microprocessor. Motorola's 680x0 processors (68030, 68040, etc., used in Apple's Macintosh line of computers), are as popular as Intel's 80x86 family (80386, 80486, etc., used in IBM's PC computers). The loss of standardization, while good for competition, will mean added trouble in robot communications.

Memory

Eight bit computers can typically address 64 kilobytes of memory. Sixteen bit computers are usually limited to 1M (megabyte), while 32 bit computers can address 4G (gigabytes) of memory. This memory is not all available for the robot user to store programs, locations, or variables. In this section we will discuss some of the other information stored in a robot's ROM and RAM memory. Figure 8.15 demonstrates how the levels of programs in memory are arranged.

ROM memory chips are supplied with the robot, and occupy some of the available memory space. *The basic input/output programs are in ROM.* These programs allow the computer to receive and send information via sensor interface boards, actuator interface boards, serial communication boards, keyboards, monitors, teach pendants, etc.

ROM also includes the **servocontrol programs,** which calculate the output required to move an individual axis to a demanded position, or at a demanded velocity, acceleration, or torque. The servocontrol program uses feedback from sensors to calculate the outstanding difference between actual and required position (or velocity, acceleration, torque). Some robots have built-in analog computers as axis servocontrollers, although fast digital computers now allow more of the servocontrol to be implemented in the digital computer. Some robot suppliers increase robot speed by providing a separate computer-on-a-card for each of the axis servocontrols. These slave computers respond to commands from the controller's master microprocessor.

Almost all robot controllers include ROM programs to allow tailoring of the whole robot's motion. These **motion control** features allow coordination between the individual axis so that, for example, all axes will start and stop simultaneously, or so that the ultimate tip velocity of the gripper is maintained constant, etc.

ROM must contain **operating system** programs. The operating system allows the user (who doesn't have to be a human) to give commands such as "run," "learn," "edit," etc.

Finally, ROM also includes the **application program** that instructs the computer how to respond to instructions in a user-written program. This application-level set of routines also includes features allowing the user to write, edit, and check a program before running it, or to translate the program instructions into a form that can be more easily interpreted by the computer. Application programs also allow the user to teach and modify positions, paths, and variable values. Not all application program features have to be included in ROM. Some parts may be loaded into RAM from external program storage (e.g., floppy disks) only when needed. Other times, the features are included in offline programming packages so that they never need to occupy space in the controller's memory.

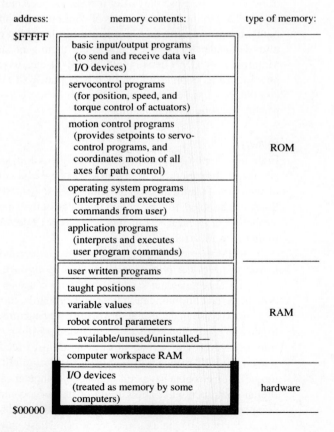

Fig. 8.15 Robot controller memory organization

The remaining addressable memory space may be fully or only partially occupied by RAM memory chips. Some suppliers include only a minimum amount of RAM and sell add-in memory as optional.

Whatever RAM is available is used to store **user programs, taught locations, current variable values, setup parameters,** and the **working data** required by the programs in ROM.

Setup parameters are used by the motion control programs. They may include values set by the user or by the user program, such as a speed setting, the distance from the gripper mounting flange to the tool center point (TCP), or other user-adjustable operating characteristics.

Input/Output Capabilities

In figure 8.16, we see that there are a variety of inputs and outputs available with robot controllers. Computer I/O was covered in more detail in the previous chapter, but we will reexamine I/O, as used by a robot controller, here.

The robot must output power to the actuators at each of the **individual axes.** This generally requires digital to analog conversion so that the digital computer can drive the (for example) DC motor or servovalve. Similarly, the servocontrol of each axis requires that the computer input either digital or analog sensory information from the sensor at each axis. These are not I/O that the user usually need be conscious of.

Teach pendants are supplied with most robots. The way in which the controller handles input from the pendant and output to the pendant display is not something the user needs to worry about. Teach pendants come in a variety of levels of complexity. Rudimentary teach pendants allow the user to control the axes of the arm manually, and to tell the controller to memorize robot positions. Simple robots can be programmed entirely

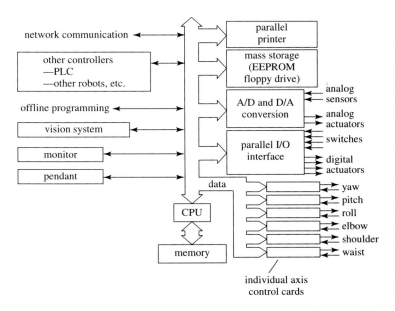

Fig. 8.16 Robot I/O devices

from the pendant, so their pendants must be more complex, including buttons for each operating system command (save, edit, run, etc.) and each application programming instruction (move, read input, close gripper, etc.).

All robots allow sensors to be connected, although the required interfaces may be options. The most basic interfaces allow switches and switched sensors (such as proximity sensors) to be connected. **Digital interfaces** also allow the robot controller to turn actuators on and off. The robot purchaser must ensure that the robot offers sufficient digital input and output points, capable of handling the required power levels.

Most robots also offer, as an option, **analog inputs and/or outputs.** Analog inputs allow transducer type sensors (such as position sensors) to be used by the robot. Analog outputs allow control of analog actuators (such as proportional servovalves). Analog I/O requires more computer memory than discrete I/O. One 16 bit data word can store the input information from 16 switches, or can store input information from (typically) one analog sensor.

Many robots offer battery-backed memory so that user programs aren't lost when power goes off. Even with these robots, important user-written programs should be saved on **floppy disks or EEPROMs** in case RAM is accidentally erased. Many controllers thus include floppy disk drives, tape drives, or EEPROM sockets.

If the robot has **serial data ports,** numerous features become available. Robot application programs may contain instructions allowing the robot to *receive data from other controllers (or intelligent peripherals* such as bar code readers), and to react to the information received. The robot may also be able to send important information (such as the results of a test performed on a workpiece) to other controllers.

A serial port also allows inclusion of the robot in a larger network of computers. The earliest application of this feature was to allow **offline programming.** Robot suppliers offer offline programming software and standard interface cables so that the robot's user programs can be created on standard desktop or mainframe computers. Offline programming sometimes allows the user to check the programs for syntax errors, and sometimes even to model the robot motion as it will occur during program execution. In modeling, a graphical image of the robot executes the user program on the video display terminal.

Programs can be downloaded from a personal computer to the robot controller via the serial link, using a **communications program** included with the offline programming software. Similarly, programs, locations, variables, and setup information from the robot controller can be uploaded from the controller to the remote computer.

Offline programming software often includes **"terminal emulation"** software that allows the user to use a personal computer as a terminal to give operating system commands to the robot controller. Offline programming software thus allows robot suppliers *not* to provide costly components such as terminals, disk drives, or EEPROM interface.

Inclusion of a robot controller in a **distributed control system** or in a **local area network** via serial links means that the robot can be placed under the control of a supervisory computer that can issue commands to the robot's operating system. This means that the supervisory computer can download programs, upload variable values, issue commands, or respond to messages from the robot. The serial port is thus the tool through which robots are integrated into larger, more flexible manufacturing systems.

8.5 LEVELS OF SOPHISTICATION

The science of robotics is improving rapidly. New features become available as fast as more powerful microprocessors become available to make them viable. Older robots, quite adequate for simple jobs, stay on the market as long as customers continue to buy them.

The major advances in robotics have been in:

- **sensor and actuator technology** (discussed in other chapters)
- **communication capabilities** between robot controllers and other computerized equipment (discussed in the previous section and the previous chapter)
- increasingly sophisticated **robot programming languages**
- improved **path control**

8.5.1 Path Control

The development of good quality servosystem controllers (analog computers) in the forties and fifties led to the first industrial robots as we know them.

Position sensors were strapped to skilled machinists, and as they worked, their movements were recorded on magnetic tape. The tapes were later played back as position control signals for electrical arms (robots) turning out the same parts as the machinists had turned out.

The disadvantage of these early robots was that, by the time they could overcome friction and inertia to respond to a changing position control signal, the control signal had changed again! The automated system could not keep up with the position control commands.

Modern robots still suffer from the effects of varying friction and inertia. With digital computer controllers, however, we can issue a position-change command, then *wait* until the sensors indicate that the robot has completed (or almost completed) the move before issuing another command.

Figure 8.17 shows how a typical robot axis responds, in terms of velocity, to a move command.

The axis is gradually accelerated. Acceleration is controlled so that it will smoothly move to maximum velocity for the middle part of the move.

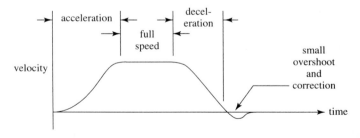

Fig. 8.17 Velocity response of a robotic axis to a move command

Axis velocity remains maximum until the axis position sensor indicates that the goal is near, so velocity gradually reduces just as it was increased. During a short move, deceleration may begin before maximum speed is reached.

Ideally, velocity would go to zero just as the axis reached its required position. In fact, in most robotic controllers, a little overshoot is allowed, then corrected for immediately, in the interest of optimizing the move time.

Some robot languages include commands so that the programmer can specify maximum velocities and accelerations.

Point to Point Robots

Most robots work quite well in **point to point** mode. Point to point moves require that only the final position for each axis be known. Point to point moves are demonstrated in figure 8.18.

The earliest digitally controlled robots could process only one axis move at a time. The axes were moved **sequentially** until each had moved from start point to end point. Modern controllers are faster, so sequential point to point path control is no longer used.

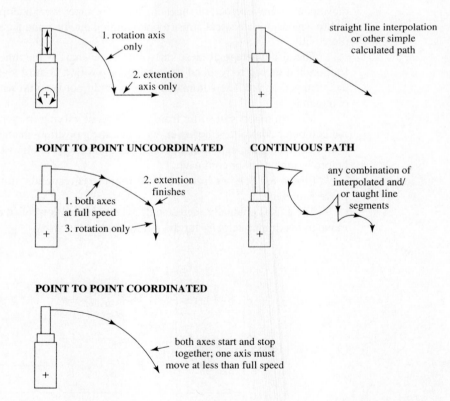

Fig. 8.18 Robot payload paths with various types of path control (cylindrical robot, top view)

When all robot axes drive simultaneously, they may move in **uncoordinated** point to point mode. When a move command is issued, all axes drive at maximum speed toward that new position. Some axes finish before others. Uncoordinated point to point path control is now rarely seen.

Coordinated point to point path control requires that all the axes start moving together and finish simultaneously. To accomplish this, it is necessary that the controller first calculate the reduced maximum velocities for all but one of the axes. A further "improvement" with coordinated point to point path control calculates acceleration so that each axis reaches its maximum speed only for an instant at the midpoint of the move, then immediately begins to decelerate. The resultant move looks very smooth, and probably improves robot sales, but it means that the robot does not complete the move as quickly as it might. The vast majority of robots on the market today offer coordinated point to point path control.

Point to point robots do not allow the user to control the path that the robot takes in moving its payload from one point to the next. Knowing that moves will not be in straight lines, the user must carefully plan to ensure that the arm will not strike an obstacle on its way to a new position. Figure 8.18 demonstrates how the payload path is affected by the type of path control. In some applications (e.g., welding), control of the path is just as important as the final destination.

Controlled Path Robots

Many inexpensive, point to point robots now offer limited **controlled path** features. In these robots, controlled path motion instructions cause the controller to calculate a series of interim points, or **vias,** between the current location and the final location. The vias are supplied sequentially to the individual axis servocontrollers. Most controlled path robots offer linear interpolation, allowing the robot to move its payload in a straight line from one point to the next. Some offer circular interpolation, for moving through defined arcs. A few offer more sophisticated path interpolation. Robots will eventually be capable of controlled path motion through any curve that a CAD system can draw.

With current controllers, calculation time noticeably slows controlled path moves. Cumulative error after long controlled path moves can be objectionable.

One benefit of controlled path robots is that they can calculate move sequences that they haven't been taught, to respond to sensory input. Robots are now available that can use information from vision systems, for example, to find and pick up randomly oriented parts from moving conveyor belts.

Continuous Path Robots

Continuous path robots can have the **"pathnodes"** in a path move entered into memory as program data, or they can be captured while having the arm manually guided through the move. Pathnodes are issued in sequence by the robot controller to the axis controllers when the path move is being executed. The robot can often be instructed to move in straight lines from node to node, or to blend the nodes into arc sections during path execution. Enormous amounts of memory can be consumed by even short programs.

Although running a continuous path program is similar to the playing back of analog tapes in early robots, the difference is that with digital playback it is possible to avoid issuing a new pathnode command until a previous pathnode move has been completed.

The challenge is to generate a smooth motion without actually stopping at each pathnode, yet to minimize deviation from the taught path.

8.5.2 Programming Languages

Most robot programming languages look similar to other high level computer programming languages. The only real difference is that robot languages include instructions such as "Move to."

Only very simple languages are needed for robot controllers with 8 bit microprocessors. Sequences of **"move"** commands, issued from a pendant, with **"ifinput"** and **"output"** type commands for external sensor and actuator use, and perhaps a **"delay"** command, are sufficient. Add instructions that allow the controller to **"goto"** other sections of the program to repeat program sections, and the ability to **count** cycles, and you have a language sufficient for simple repetitive tasks.

Some languages for 8 bit controllers, but more frequently for 16 bit controllers, allow selection of robot **speed,** and allow the inclusion of **subroutines.**

Some programming languages at this level allow the programmer to specify the **precision** with which the robot is to perform its moves. In the interest of speed and smoothness of the move, most robots do not usually finish a move or come to a complete stop before beginning the next. They begin the next move as soon as they come near the first destination. The precision command sets the amount of allowable positional error. Alternately, some programming languages allow the user to specify that certain moves must be **finished** before another move can be initiated.

Sixteen and 32 bit controllers allow more sophistication in programming languages. The programs may allow **interrupts.** Interrupt instructions tell the controller to constantly monitor specified input conditions (sensors, signals from supervisory computers, etc.), and to jump to a subroutine or to a separate program as soon as the anticipated signal is received. Use of instructions setting up (or suspending) interrupt conditions means that the program does not have to include the frequent sensor checking instructions that would otherwise be necessary.

These more powerful controllers also allow programs to include motion control commands such as those limiting **acceleration,** or those changing the **gains in the servosystem** routines. **Linear interpolation** commands are increasingly common in 16 bit controllers, and 32 bit controllers often allow **circular interpolation** and other specifiers of path shape. **Trigger** commands allow the program to signal when the robot has moved through specified points in space.

Languages written for 16 and 32 bit controllers also may include the ability to **import, evaluate, manipulate, and output strings of ASCII information.** Some obvious applications exist — for example, where bar codes are used on workpieces and tools in the robotic workcell.

Early robotic programming languages were very similar to the languages used to program NC machinery: a series of **binary machine instructions** or **alphanumeric codes.** Slightly more sophisticated programming languages look and operate like **BASIC.** They provide a sequence of English-like instructions, which the controller **interprets** one at a time, executing each instruction before examining the next. For increased speed, entire user programs can be **compiled** (translated from English-like instructions to machine language) and the compiled version saved for execution by the controller. These programming languages are often based on computer programming languages such as FORTRAN, Pascal, or C.

More recently, attempts to increase the flexibility of robots has lead to attempts to use **artificial intelligence** programming languages, so that robots can evaluate unpredictable sensory information, or ambiguous operator instructions, and make decisions of their own. This level of programming sophistication is still highly experimental. Given that robots are being used now for precision brain surgery requiring more manual dexterity than humans possess, and given that artificial intelligence "expert systems" are now used to diagnose patient illnesses, some interesting potential combinations of applications already exist.

8.6 SUMMARY

Robots, often used as sophisticated actuators, are actually **complete automated systems** including actuators, sensors, and controllers. Some **pick and place robots** do not include the controller, and often not the sensors, but true robots require both. In a true robot, each individually controlled axis is a complete **servosystem.**

The robotic **arm** has several axes, arranged such that the three major degrees of freedom allow us to classify robots as **articulated** (all rotary axes), **spherical** (two rotary axes), **cylindrical** (two translational axes), **cartesian** (all translational axes), or **SCARA** (three rotary axes, each with a vertical axis of rotation). Robots, except the SCARA, are usually provided with at least two of the **wrist roll, pitch, or yaw** rotary degrees of freedom. The gripper is additional and is often custom designed for a particular job. SCARA robots require a **Z axis** degree of freedom.

Robots are usually powered by **electrical motors,** which are quiet and clean. For more power and speed, **hydraulically actuated** robots are available. **Pneumatic power** is sufficient for most pick and place robots.

A robot controller is a full computer system. It has a **processor,** an **operating system in ROM,** and sufficient **I/O capabilities** to handle the arm's sensors and motors, the teach pendant, and perhaps some mass storage device such as an EEPROM, or to communicate with an external computer. Modern robots have 16 or 32 bit **microprocessors,** allowing sufficient memory for **coordinated motion control** programs, and often for other adjustment-of-motion commands, such as "precision" commands, which control how the robot blends sequential move series. Many robots offer **controlled motion** features, so that the programmer can specify the shape of the path that the tool endpoint describes as it moves from point to point. With more memory capability, **continuous path** moves, where each move consists of many interim points that must be moved through, is possible.

The more powerful the robot controller, the more powerful the programming language can be. Modern languages allow **subroutines, interrupt service routines,** and the **import and manipulation of data from smart peripherals.** Work is being done to enhance the ability of robots to make decisions by using artificial intelligence (AI) programming.

REVIEW QUESTIONS

1. Why are pick and place robots identified in this chapter as "near-robots"?
2. How many linear and how many rotational axes are there in spherical robot geometries? Cylindrical?
3. What do the letters SCARA stand for and what do they mean?
4. On which major geometry of robot would you avoid putting a wrist with pitch and yaw?
5. Why are open loop controllers not good robot controllers?
6. Why are pneumatic servo-robots no longer commonly available?
7. What are the two preferred types of electrical motors for robotic actuators?
8. Why can the position sensors on a robot not precisely indicate the location of the end of the gripper?
9. How does an angular gripper differ from a parallel action gripper?
10. What type of end effector would be best for picking up large, flat sheets of glass? Why?
11. Identify three different ways that an electrical power supply might have to condition electrical power for robot use.
12. Which component is at the heart of a robot controller?
13. Identify one function of each of the following programs in ROM:
    a) Basic input/output system
    b) Servosystem
    c) Motion control
    d) Operating system
14. Why do almost all robot controllers offer digital I/O capability?
15. Why do most robot path control programs allow a small overshoot in each move?
16. Which requires more RAM memory and which requires a more complex path control program: continuous path control or controlled path control?

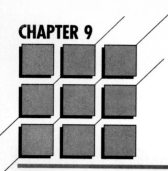

CHAPTER 9

THE AUTOMATED WORKCELL AND ITS FUTURE

CHAPTER OBJECTIVES

After completing this chapter, the reader will be able to:

- Define "workcell."
- Describe the operation of controllers and other components in a typical workcell.
- Identify the signal exchange and data exchange requirements for controllers in workcells.
- Describe the difference between workcells as:
  - islands of automation
  - parts of Flexible Manufacturing Systems
  - parts of Computer Integrated Manufacturing systems
- Identify the data-exchange requirements in a Flexible Manufacturing System.
- Identify and briefly describe the software components of a Computer Integrated Manufacturing system for:
  - management decision support
  - marketing decision support
  - computer aided design and testing
  - computer aided process planning
  - manufacturing and inventory control
  - data/information availability

9.0 THE AUTOMATED WORKCELL AND ITS FUTURE

In automated workcells, many components work together, under the control of one or more controllers. Automated workcells are often combined into automated systems such as that shown in figure 9.1, and automated systems can be combined into automated factories.

Fig. 9.1 A flexible, palletized conveyor assembly system. (Photograph by permission, Automated Tooling Systems Inc., Kitchener, Ontario, Canada.)

Some automated workcells can have more than one purpose. If the workcell can change its own function, it is said to be part of a **Flexible Manufacturing System (FMS)**.

If a workcell is fully automated, controllers within the workcell must exchange information with each other. If a workcell is to be more than an isolated **island of automation,** then it must communicate with controllers in neighboring workcells.

The "future" in automation approaches fast. Advances in computer technology and advances in software capabilities occur at a dizzying rate. In fact, to some extent, the future is now. There are many manufacturing plants where major portions of the manufacturing occur without human intervention! Some manufacturing plants have been described as "lights out" plants because, without humans, the manufacturing can occur just as well in the dark as in a lighted plant!

Computer control can mean more than just control of manufacturing. All corporate activity is now being analyzed and computerized. Computers can, and do, accept orders from customers' computers. They decide what is going to be manufactured and when. They decide when and how to store and ship goods.

Automation to this level requires careful programming of controllers so that they exchange the data that is required, when it is required. When computers work with each other to control manufacturing, the system that they are part of is said to be a **Computer Integrated Manufacturing (CIM)** system.

9.1 THE WORKCELL — AN EXAMPLE

Figure 9.2 illustrates how several components and controllers can be combined into a workcell.

What the Workcell Does ...

This workcell:

> 1) Assembles connectors onto circuit boards.
>
> Nearly-completed circuit boards arrive at the workcell on a belt conveyor. A PLC-controlled fixture at the end of the conveyor orients them, one at a time, so that the robot can pick them up.
>
> The robot moves the circuit board from the conveyor to the soldering fixture.
>
> A vibratory bowl feeder, under PLC control, separates and orients connectors, which start loosely-jumbled in the feeder bowl. As the vibratory bowl outputs oriented connectors, a fixture precisely positions them so that the robot can pick them up.
>
> The robot picks up a connector and places it on the board in the soldering fixture.
>
> A soldering head at the soldering fixture permanently attaches the connector to the circuit board.
>
> 2) Tests the resultant assembly.
>
> The robot moves the assembled and soldered circuit board to the testing fixture.

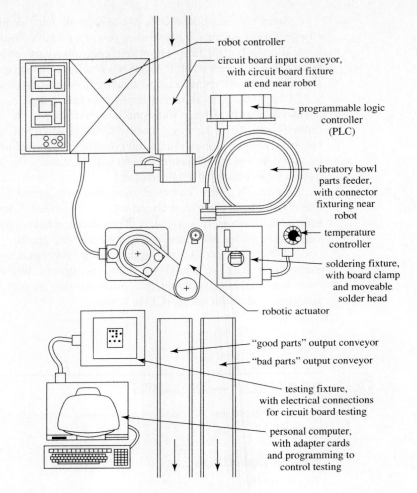

Fig. 9.2 An automated workcell

The tester performs electrical testing of each component on the circuit board, and tests to ensure that the complete circuit works properly.

3) The robot receives a "test passed" or a "test failed" message from the tester, then either:

places working circuit boards on a "good parts" output conveyor, or

places defective circuit boards onto the "bad parts" output conveyor.

What Each Controller Does ...

There are four controllers in this workcell. Figure 9.3 shows how they are interconnected within the workcell. Each controller has its own program and runs independently, except that each program periodically requires signals from other controllers, or sends signals to other controllers.

To improve readability, the following description of the workcell operation has been simplified. If this description is used as a model to build and program a similar workcell, the reader will find that more detail is required in some places, and that some mechanisms and programs could be better designed. A couple of such areas will be identified in the following explanation. The simplification is intentional: to allow a wide range of controllers, production equipment, and communication functions to be presented in a few short pages.

The first of the four controllers is a single **programmable controller (PLC),** which monitors the sensors and controls the actuators at the vibratory bowl and at the input belt conveyor. (See chapter 7 for programmable controllers.) Figure 9.4 illustrates the PLC, which consists of:

- a central processing unit (CPU)
- a power supply module
- a module to read digital (on/off) input signals
- a module to output digital signals

Figure 9.4 also shows the sensors and the actuators at the circuit board fixture and at the vibratory bowl feeder output. Capacitive proximity sensors detect the presence of circuit cards or connectors in the fixtures. When presence is detected, the normally-open sensors output a 24 VDC signal to the PLC; 24 VDC signals from the PLC actuate the

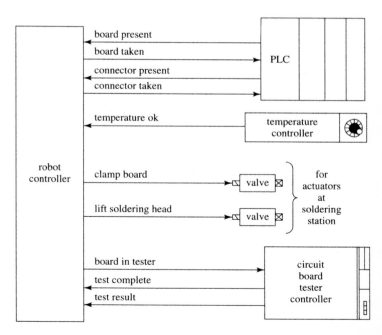

Fig. 9.3 Interconnections between controllers in the circuit board assembly workcell

solenoid valves. The actuated solenoid valves provide air to extend the air cylinders, which precisely shuttle the circuit board or connector sideways.

Figure 9.4 also includes a description of the digital signals exchanged between the robot and the PLC, and a simplified copy of the ladder logic program that the PLC would run. Following is a description of each of the six segments in the PLC program.

Segment One monitors the sensor at the end of the input conveyor (I1). When a circuit board is detected, the PLC actuates the board fixture (Q1). Boards arriving on the conveyor are not identically oriented each time. The fixturing is required to position each board exactly in the orientation that the robot requires to pick the board up.

The board fixture output is "latched," which means it stays on until another program segment "unlatches" it.

Segment Two monitors a signal line from the robot (I4) for a signal meaning that the robot has removed the board from the board fixture. Once this signal is received, the PLC "unlatches" the fixture actuator (Q1), returning the board fixture to its original position so that another board can enter.

Segment Three sends a "board present" signal (Q4) to the robot while a circuit board is held (Q1) in the board fixture. Since this signal is not latched, the PLC always turns this signal off unless the fixture is actuated.

Segment Four monitors a sensor (I2) at the output of the vibratory bowl feeder. When a connector is detected, the PLC actuates (latches) a connector fixture (Q2).

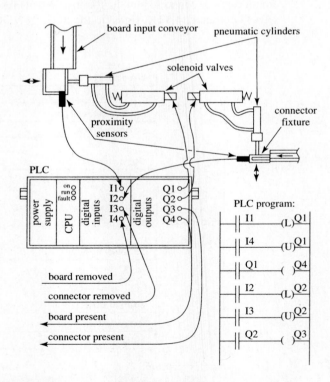

Fig. 9.4 PLC in the circuit card assembly workcell

Segment Five monitors a signal line for the "connector removed" signal from the robot (I3). When this signal is received, the PLC unlatches the fixture output (Q2) to allow another connector to enter it.

Segment Six sends a "connector present" signal (Q3) while a connector is held (Q2) in the connector fixture.

The second of the four controllers is the **robot's own controller** (see chapter 8). In this example, the robot controller is the workcell's master controller. All the other controllers exchange signals with the robot controller, but not with each other. Figure 9.5 illustrates the sequence of the robot's duties, including the information exchange duties. The robot continuously repeats the following cycle:

1. The robot waits for a signal from the PLC indicating that a **circuit board** has arrived, then picks up the circuit board from the circuit board fixture. As it removes the board, it briefly sends a signal to the PLC to indicate that it has done so. The robot places the board in the soldering fixture.

2. The robot waits for a signal from the PLC indicating that a **connector** has arrived, then picks up the connector from the connector fixture. As it removes the connector, it briefly sends a signal to the PLC to indicate "connector removed." The robot places the connector onto the board in the soldering fixture.

3. If the soldering controller **IS NOT** sending the "temperature OK" signal, the robot waits while the soldering head heats up. If the soldering controller **IS** sending the "temperature OK" signal, the robot activates the clamp at the soldering fixture. One half-second later (to ensure clamping is complete), the robot controller actuates the lift that brings the heated soldering head into contact with the (pre-soldered) connector contacts. (The reader will recognize that this could be improved.) After delaying for 5 seconds to ensure the solder has melted and flowed to connect the board to the connector, the robot controller lowers the soldering head, and unclamps the board. The robot moves the circuit board to the circuit tester.

4. The robot signals the tester that a circuit board is in the tester, then waits for the "test complete" signal.

5. The robot reads the "test result" signal line from the tester. If this signal is on, the board has passed the test, so the robot moves the board to the "good parts" output conveyor. If this signal is off, the board has failed the test, so the robot moves the board to the "bad parts" output conveyor.

6. The robot signals the tester that the board has been removed.

The third of the four controllers is a **differential gap on/off controller** (see chapter 6) at the soldering fixture.

As illustrated in figure 9.6, this controller performs the following actions.

1. The controller receives a temperature setting via a manually adjusted potentiometer dial. This setting specifies the temperature that the soldering head must reach at the beginning of each working day and must hold during the working day.

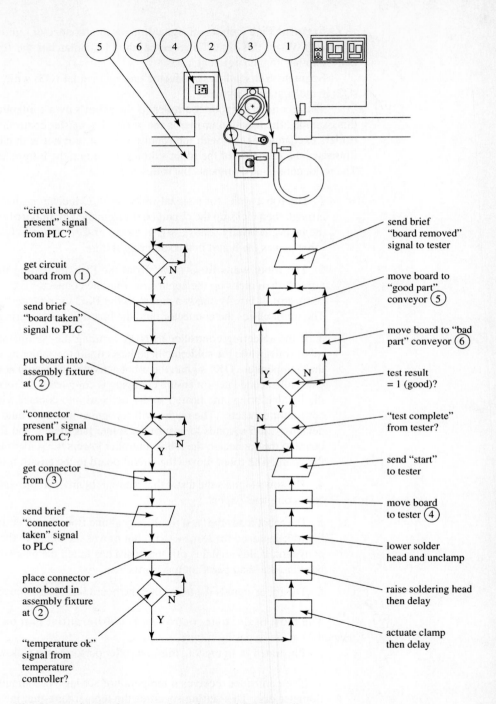

Fig 9.5 Robot in the circuit board assembly workcell

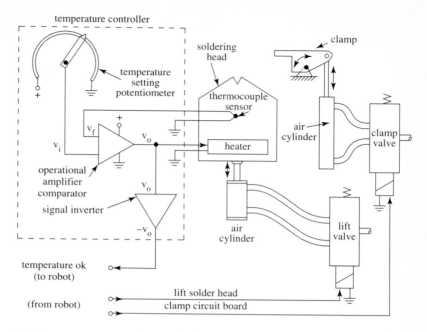

Fig. 9.6 Temperature controller and soldering fixture

Another potentiometer (not shown) inside the controller specifies an "allowable range" of deviation from the temperature setting.

2. The controller receives a feedback signal from a temperature sensor inside the soldering head. The sensor is a thermocouple. It outputs a voltage (roughly) proportional to the temperature sensed.

3. The controller compares the temperature setting with the sensed actual temperature. If the actual temperature is less than the low end of the allowable temperature range, the heater is turned on. If the actual temperature is above the upper allowable temperature, the heater is turned off.

4. The controller then outputs a signal meaning "temperature OK" when the heater is off. (The reader will recognize that this could be improved.)

There are clamping and soldering head positioning actuators at this fixture, but they are not controlled by the temperature controller. They are controlled by the robot.

In this example, the testing of the completed assembly is complex enough that **the tester requires its own controller.** This is the fourth controller. The tester is a mechanical device that contains electrical probes and contacts. The tester controller is a personal computer with special interface cards installed to enable it to exchange DC signals with the robot and control the actuators in the tester, and to perform assorted digital and analog testing of the circuit board.

Figure 9.7 illustrates the tester and the tester control program. (The program listing is intended for readability, not for accuracy in any programming language.) The tester performs the following actions.

1. The tester waits for a "start test" signal from the robot.

2. The tester then turns on a vacuum to draw the circuit board down against a bed of contact probes, and inserts a plug into the connector on the circuit board.

3. The tester opens a new file on the hard disk. The filename identifies the board. Files will contain information about boards that fail testing. Repair personnel will use the contents of these files in repairing faulty boards.

4. The tester sets the variable "RESULT" equal to one.

5. Via the contact probes, the controller tests the components on the board, one at a time. If a component fails a test, the results of that test are written into the file on hard disk, and the variable "RESULT" is set equal to zero.

6. The tester uses the plug in the connector to provide the board with signals typical of those it will receive in later use. If the board fails to perform as it should, the test results are added to the file on the hard disk and the variable "RESULT" is set equal to zero.

7. The tester outputs the value of the variable "RESULT" on one output line, and signals "done" to the robot on another output line, when the testing is complete.

9.2 ISLANDS OF AUTOMATION

The circuit board assembly workcell, described in the previous section, operates in isolation from the rest of the manufacturing plant. Since it does not exchange information with other workcells, it is considered to be an "island of automation."

Most manufacturing companies begin to automate by creating islands of automation. The islands are in fact experiments in automation technology. Unless a company chooses to hire experienced automation designers, it must gain its own experience by a combination of whatever training it can find, and by practical experimentation. In other words, it is cheaper to learn by making mistakes in the design of inexpensive islands of automation, than to learn by making mistakes in the design of large, expensive automated systems. Despite the scorn islands of automation receive in literature about Computer Integrated Manufacturing (CIM), *creating islands of automation is an essential step in the move toward well-designed CIM.*

Once a successful island of automation has been built, it is time to move on. Subsequent automated workcells should be designed so that they can become integrated components of a production line. Eventually, all automated workcells should become **integrated components** in the corporation's worldwide manufacturing capability.

The workcell controller in figure 9.2 should, at the very least, know whether there is room on the "good parts" output conveyor before it tries to place a good circuit board

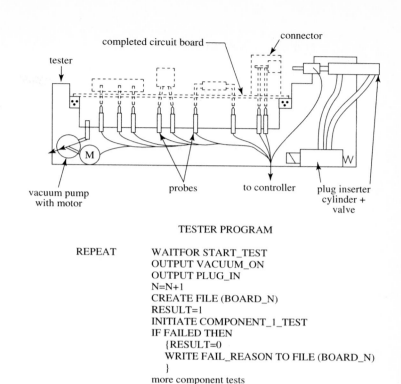

```
                         TESTER PROGRAM

         REPEAT        WAITFOR START_TEST
                       OUTPUT VACUUM_ON
                       OUTPUT PLUG_IN
                       N=N+1
                       CREATE FILE (BOARD_N)
                       RESULT=1
                       INITIATE COMPONENT_1_TEST
                       IF FAILED THEN
                          {RESULT=0
                           WRITE FAIL_REASON TO FILE (BOARD_N)
                          }
                       more component tests
                                  ⋮
                       INITIATE BOARD_FUNCTION_TEST
                       IF FAILED THEN
                          {RESULT=0
                           WRITE FAIL_REASON TO FILE (BOARD_N)
                          }
                       CLOSE FILE (BOARD_N)
                       OUTPUT PLUG_OUT
                       OUTPUT VACUUM-OFF
                       OUTPUT RESULT
                       OUTPUT TEST_COMPLETE
                       WAITFOR BOARD_REMOVED
                       OUTPUT -TEST_COMPLETE

                       GOTO REPEAT
```

Fig. 9.7 Circuit card testing in the assembly workcell

there! In a fully automated manufacturing plant, this workcell might receive information about quality problems encountered at other workcells and respond by performing different circuit board tests. If this workcell was designed differently, it might be able to assemble various different circuit cards, depending on the orders the plant has received in the last half hour!

9.3 FLEXIBLE MANUFACTURING SYSTEMS

A manufacturing production line that *can vary the way it works, or the products it produces,* is a Flexible Manufacturing System (FMS). To be a true FMS, the production line should be able to modify itself without requiring a human to help. The modification might be in response to data received from other controllers, or in response to conditions sensed by sensors.

Numerically controlled (NC) **machining centers** are excellent components for flexible manufacturing systems. They can change their own cutting tools to machine different shapes, can exchange information with other controllers, and can change their sequence of operations in response to new information received.

9.3.1 The FMS Workcell — An Example

If the circuit board assembly workcell shown in figure 9.2 were modified to become a part of an FMS, it might look like figure 9.8. This FMS workcell performs the following tasks.

1. The workcell accepts a variety of circuit cards (all received on the same input conveyor, in no particular order). There is a **bar code reader** on the input conveyor to identify the incoming circuit card type.

2. Depending on the type of circuit card that has arrived, the workcell selects the correct connector or connectors from a variety of vibratory bowl feeders at the workcell.

3. The workcell places and solders connectors in the appropriate locations for the board type.

4. The workcell runs various tests of circuit card components and circuit performance, depending on the type of board being assembled.

One important change is the addition of another controller, called a **cell controller.** The cell controller is required only if the existing controllers in the workcell cannot handle the extra programming and data requirements required to make the workcell "flexible."

What the FMS Workcell Does ...

As a circuit board approaches the workcell, the bar code reader reads the bar code printed on the part. When the circuit card trips the light beam sensor at the end of its conveyor, the cell controller requests the part number from the bar code reader. The bar code reader responds by sending the circuit card's part number (by asynchronous serial ASCII transmission) to the cell controller.

For each possible part number, the cell controller is programmed to initiate the appropriate programs at the other controllers in the workcell. In this example, the **PLC**'s program needs only to be expanded to allow it to control more fixtures.

The **robot** can store several different programs in its controller. The cell controller must send it a serial communication telling it which program to run. The robot still exchanges simple on/off signals with the PLC, solder head actuators, temperature controller, and tester.

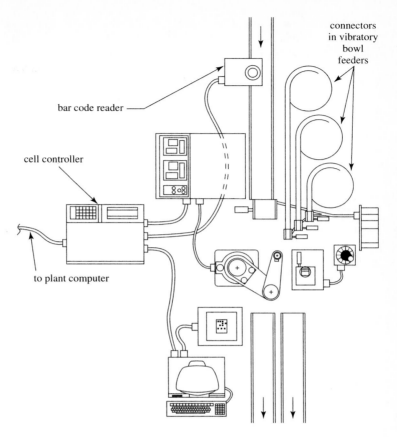

Fig. 9.8 A Flexible Manufacturing System workcell for circuit card assembly

The **tester** is controlled similarly to the robot. It has several programs in its memory, and the cell controller tells it which program to run.

The robot and the tester controllers report "done" to the cell controller when they have completed their tasks, or report the reasons why a task cannot be completed. The tester controller communicates test results to the cell controller.

The cell controller is programmed to signal the plant computer if the assembly task cannot be completed for any reason, or if an unusual number of defective circuit cards are detected at testing. The plant computer is programmed to alert maintenance personnel when these production problems are detected.

In the above example of a flexible circuit board manufacturing workcell, the *workcell automatically varies its sequence of operations* dependent on the bar code that is read from the incoming circuit board. In another type of flexible manufacturing system, a central computer tracks assemblies and components as they pass through the FMS. The circuit board assembly workcell would receive *a message from the central plant computer telling it which production sequence* to begin. The bar code reader wouldn't be necessary.

If there are many different products or production sequences, then the *cell controller might have to store programs for the other controllers* and supply them to the other controllers as they are needed. More commonly, the cell controller only handles communications between the plant computer and the workcell controllers. *The plant controller "downloads" (sends) robot programs, tester programs, and PLC programs to the cell controller* as they are needed. The cell controller receives them and forwards them to the workcell controllers. Test results are "uploaded" to the plant controller by the cell controller.

Many manufacturing plants in North America have at least small flexible manufacturing systems.

9.4 COMPUTER INTEGRATED MANUFACTURING (CIM)

In a Computer Integrated Manufacturing (CIM) facility, *computers handle the scheduling of manufacturing, in addition to controlling the equipment that does the manufacturing.* There are very few true CIM facilities in North America or anywhere else (so far).

We briefly examined the components of a Computer Integrated Manufacturing plant in chapter 1, Introduction to Automation. The diagram that we examined in that chapter is reproduced here as figure 9.9 and is explained in a little more detail.

A CIM environment such as that in figure 9.9 *may include workcells similar to the circuit card FMS workcell* already examined in this chapter. It also may include NC equipment, under the control of cell controllers.

A CIM environment *usually requires that one computer be designated the central or "host" computer.*

The host computer may be the manufacturing computer, or may be any other large computer in the CIM system.

One or more **local area networks (LANs)** are required to connect the host computer to other computers in the manufacturing plant and in the front office; to each cell controller; and to other controllers, personal computers, or terminals. A **gateway** is needed to allow incompatible computer systems and LANs to communicate. Gateways are computers that translate messages from one form to another form, without altering message contents.

A CIM system often requires that computers at several different locations be linked. Satellite links, as shown in figure 9.9, are used in some **wide area networks (WANs).**

9.4.1 CIM — An Example

A Computer Integrated Manufacturing environment of the future might operate as the following paragraphs describe. The elements that are described in the following section all exist! All are in use somewhere in industry, although not all in any one system. Several of these elements need improvements before they can be used as easily and successfully as the following description suggests.

1) The first step in a product's life cycle probably can't be automated or computerized. The need for a new product must be recognized. Once the need is recognized, however, the description of the *proposed product is electronically transmitted* to a manufacturing company.

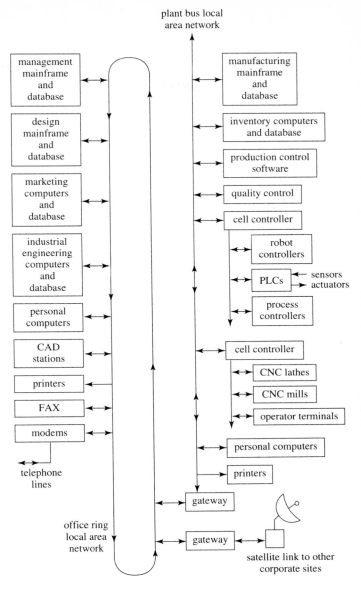

Fig. 9.9 A Computer Integrated Manufacturing environment

2) Company management must *determine if the proposed product fits with the company's philosophy* about what it should be producing. Computer-based **decision support software (DSS) programs** help in this decision. If the original product proposal is in the appropriate data form, the DSS program uses that data as input.

3) Marketing uses another **decision support software (DSS) program** *to estimate the probability of being able to sell* the proposed product. The DSS also

9.4 Computer Integrated Manufacturing (CIM) **269**

recommends the best marketing approach, and suggests whether the product should be manufactured primarily for low cost or for high quality.

4) Now the design department gets its turn. **Computer Aided Design (CAD)** software is used to generate *part drawings and bills of materials* for assemblies.

As part numbers are automatically assigned, the **group technology (GT) software** in the **database management system (DBMS)** compares each new part with existing parts, and *suggests that the designer use existing parts if similarities are found.*

Software add-ins to the CAD software are used to *simulate testing* of parts and assemblies even before they are built, finding design weaknesses. Other CAD system add-ins *evaluate the producibility* of the designed parts, given the resources in the manufacturing plant, and *estimate production costs.*

5) In the manufacturing department, the CAD database describing the parts and assemblies are used to *develop production plans,* using computer programs called **computer-aided production planners (CAPP).** Production plans list the steps required to manufacture the product, using available manufacturing capabilities. The computer-aided production planning includes *computer generation of NC programs or other machine programs* that are required.

6) The sales department will now find that the designed (but never built) product is included in the *computer-generated list of products* the company produces. This list is immediately available to any customer with a computer link to this plant. A customer can use his *computer to request a price quote* for a quantity of the new product.

Once a customer's inquiry is received via a wide area computer network (WAN), the **Manufacturing Resource Planning (MRP II)** software in the host computer *estimates cost and delivery.* This information is sent back to the prospective customer's computer so the customer can decide whether to *place an order electronically.*

7) Once an order is received, the manufacturing department must make the product. **MRP II software** includes the newly-ordered products in the **Master Production Schedule (MPS),** and *recalculates material, labor, and tooling requirements* for all outstanding orders, including this new one.

The **MRP II software** *issues electronic purchase orders* to buy materials from other companies (verifying delivery dates with their computers). The **Shop Floor Control (SFC)** component of the MRP II program *sends manufacturing programs* to manufacturing equipment, as required. (Manufacturing equipment includes material-moving equipment as well as stationary workcell equipment.) If there are any humans in the manufacturing plant, the computer will provide them with **work schedules** via their computer terminals.

8) Computers control *automated equipment maintenance* and *monitor for impending failure* to reduce lost manufacturing time due to breakdown.

CIM cannot, however, make equipment breakdown obsolete, so the MRP II software occasionally receives messages from manufacturing equipment (or from humans) indicating that manufacturing has fallen behind schedule. The *MRP II software adjusts its production scheduling* to catch up using other available production equipment.

9) Stockroom and shipping have also been automated. As parts and assemblies are completed they are transported to the **automated storage and retrieval system (AS/RS).** The AS/RS, a computer-controlled, automated, material-transportation-and-storage system, replaces the obsolete storeroom. The computer keeps track of where everything is stored in the AS/RS, how much of each item is left, and the number and location of items that are "reserved" for outstanding orders.

When a customer's order is completed, it is *automatically drawn* from the AS/RS and *boxed, labelled, and addressed using computer-generated labels.* If several orders are to be sent to a local distribution center instead of directly to the customer, they are packaged together for cheaper bulk shipping. (Even post offices use automated sorting and material moving equipment. A postal plant in Toronto has equipment that lifts whole transport trucks to dump them.)

10) The accounting department gets some computer help, too. *Customer billing is printed by computer* when the order is shipped. The invoices (bills) are verified and sent by **humans,** however. Some things are too important to trust to computers!

9.4.2 A Word from the Real World

The potential for CIM is mind-boggling! As stated at the beginning of this section, all the described components exist and are being used. Why, then, are typical CIM implementations far less inclusive than the system described? There are several reasons, but perhaps the greatest impediment to the full implementation of CIM is: the large amount of training required before it can be *used*! Organizations contemplating CIM implementation aren't going to invest time or money unless they are confident the results will justify the effort.

The next chapter discusses how automation proposals (from the first PLC-controlled workcell to the fully automated "lights-out" manufacturing plant) can be developed, evaluated, and implemented, and how to continue to improve installed systems.

9.5 SUMMARY

Workcells are integrated systems of controllers, sensors, and actuators. Individual **controllers,** such as PLCs, robot controllers, special-purpose controllers, and personal computers, all have their own tasks in a workcell. To ensure that the workcell controllers work together, they must be **connected via signal lines, and must be programmed to exchange signals.**

Some automated workcells are designed to be able to do more than one type of task. If the workcell has a controller acting as a cell controller, it is possible to integrate the workcell into a **Flexible Manufacturing System (FMS).** In an FMS, workcells can respond to changing manufacturing conditions. An FMS **often requires a host computer** to provide data and programs to workcells as manufacturing conditions change. Data exchange is often via **local area networks (LANs).**

A **Computer Integrated Manufacturing (CIM)** system requires even more data exchange between controllers than in an FMS. Computers and software *help all corporate departments* from top administration, to design engineering, manufacturing engineering, manufacturing operations, sales, purchasing, stockroom, accounting, and even maintenance. Corporate computers maintain data for these functions and ensure that all departments use the same data.

REVIEW QUESTIONS

1. What makes a group of sensors, actuators, and controllers a "workcell"?

2. Sketch a proposed layout with the same purpose as that in figure 9.2 (assembling and testing circuit boards), but which does not include a servo-robot. Select actuators, sensors, and controllers using information from other chapters in this text.

   a) Describe the sequence of operations in the workcell.

   b) Sketch the signal exchange requirements between controllers.

   c) Describe the sequence of control functions for each controller. Be sure to include each sensor and actuator signal, and the signals exchanged between controllers.

3. How does a workcell that is part of a Flexible Manufacturing System differ from an island of automation?

4. What additional components and data exchange would be required if the layout described in your answer to question 2 was to be part of a Flexible Manufacturing System. Use some method other than bar code reading to recognize the changing manufacturing requirements.

5. Refer to chapter 7. Use the information from that chapter to describe the characteristics of a local area network appropriate for use in a Computer Integrated Manufacturing system such as that described in this chapter. Include network communication hardware and architecture. What type of information would be stored in each of the databases shown in figure 9.9?

CHAPTER 10

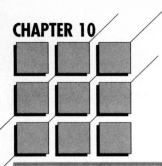

AUTOMATION: WHEN? HOW? AND OTHER NON-TECHNICAL ISSUES

CHAPTER OBJECTIVES

After completing this chapter, the reader will be able to:

- Recognize when an organization is prepared to consider automation.
- Discuss some methods of simplification and control of the manufacturing processes in preparation for automation, including:
  - Just in Time (JIT) methods
  - Statistical Process Control (SPC) methods
- Evaluate and discuss the safety requirements for automated processes, and identify some methods of including safety in a design.
- Calculate the payback period for automated systems, using a traditional approach and including costs and benefits of quality.
- Discuss current trends in societal effects of industrial automation.
- Discuss how society should react to control the effects of industrial automation.

AUTOMATE, EMIGRATE, LEGISLATE, OR EVAPORATE

Chapter 1 began this way, too. From that ominous opening, chapter 1 and the rest of the book moved into a technical discussion about automation. This final chapter returns to examine the "automate … or evaporate" question, and some other non-technical aspects of automation, including:

- when to automate
- how to take an organizational approach to automation
- some safety considerations
- cost justification for automation
- some social considerations

 ## WHEN TO AUTOMATE

Automation, chapter 1 noted, is not a magic solution to financial problems. Automation can be a valuable tool used to improve product quality. Improving product quality, in turn, can result in lower costs, and can lead to improved sales, which should help any organization's financial problems.

The author has designed automated equipment for (among others) a major telecommunications company. That company has had generally good experiences automating. They began early and built slowly. They have made mistakes, but so far the benefits have outweighed the costs. Sales of products from their automated lines have been strong, and production costs are way down.

Other companies have had less success automating. Changing large, traditional manufacturing systems to automated systems can be difficult, especially if the companies have neglected their production systems and production personnel for years. Quite a number of North American manufacturing facilities fit this description.

When should an organization automate?

Automation should be considered when it can help the organization to be profitable, within the limits of the organization's philosophy.

By "within the limits of the organization's philosophy," we mean that most companies see themselves as having a place in society, just as individuals do. While an organization can't continue to exist without profit, there are philosophical limits on how an organization will behave in its pursuit of profit. Dumping toxins into the environment is unacceptable behavior, even if it lowers production costs. Installing automation solely to enable laying off the workforce should also be considered unacceptable behavior.

In Japan, Toyota Motor Company employees are encouraged to recommend cost-cutting suggestions. Sometimes a suggestion results in one less worker required at a workcell. When this happens, a deserving worker from the workcell is promoted instead of laying anyone off! Since the suggestion was a cost saver, the company has the money to afford the promotion, and the promotion is an incentive to others.

One difficulty with the statement "Automation ... when it can help the organization to be profitable, ..." is that it already requires an organization to understand the costs and the benefits of current automation technology. Before considering automation, an organization should hire employees with automation knowledge if possible, and should provide professional development opportunities to enable existing employees to understand modern manufacturing processes. Otherwise, the organization may miss cost-saving opportunities.

"Professional development opportunities" do not have to be limited to **courses and seminars** that are taken away from the place of employment. In fact, external courses limit the exposure to only a few employees, and not necessarily to the employees who should be exposed. Being away from the place of employment discourages connecting new knowledge with old methods "back at work." Some colleges, universities, and private consultants offer training **courses delivered at the employer's site.** Larger groups of employees can share the experience, and employees are more likely to link the knowledge with their company.

An excellent way of keeping employees aware of automation capabilities is to **encourage experimentation.** Once an experimental pilot workcell has been built, programmed, and tried by in-house personnel, the practical experience will always be remembered. Employees should not, of course, be discouraged if their first efforts fail, nor should they be pushed to take on projects beyond their capabilities.

10.2 AN ORGANIZATIONAL APPROACH TO AUTOMATION

When a careful examination of the corporate goals and of customers' needs reveals that a company's products and marketing strategies are satisfactory, but that improvements in the manufacturing process are required, then the following steps should be taken:

- simplification
- automation
- computer integration

The three steps can be executed in either of two ways:

1. If the need is desperate, a **revolutionary** approach might be used. Task forces can be formed and experts might be brought in to design a manufacturing system from scratch. The resulting system can be built and employees trained or hired to staff it. If the organization is lucky, the inevitable mistakes in the design won't cause bankruptcy.

2. If the organization is lucky enough to recognize the need for change early, it can take a whole-organization approach. All its employees can help to **evolve** toward an improved manufacturing system. Mistakes will still be made, but each mistake will be on a smaller, more affordable scale. Mistakes actually become valuable learning experiences.

Whether using the task force (revolutionary) approach or the organization-wide (evolutionary) approach, the same three steps are necessary: simplification, automation, and computer integration.

10.2.1 Simplification

The current manufacturing process should be reexamined for its ability to efficiently build a desirable, quality product. Simplification of the manufacturing process is probably possible. As the process is simplified, goods will be produced less expensively, and tighter quality control will be possible.

In fact, it's usually impossible to separate simplification of the manufacturing process from improvement in product quality. An understanding of how to build quality into a product results in better design of the manufacturing process. An understanding of the manufacturing process results in improvements in the finished quality of the product.

The "simplification" step should *not* include the introduction of automation. It might even make automation unnecessary!

JIT for Simplification

Just in Time (JIT) manufacturing, as practiced by the Toyota Motor Company in Japan,[1] requires product and process simplification. As we interpret Toyota's system, there are four principles embodied in the JIT manufacturing philosophy.

The first principle is: ***Nothing should be made unless it is required.*** JIT uses a production pulling system, rather than North America's production push. Instead of manufacturing parts to maintain stock (where their presence seems to demand that they be used to produce higher assemblies), JIT systems produce only parts that are needed for higher assemblies or for direct sale.

Each production workcell in a JIT system does maintain a few finished parts, at the workcell. Other workcells or sales centers can take these parts if they have a *"withdrawal Kanban."* The number of withdrawal Kanbans is kept as small as possible so that the workcell that makes the parts doesn't need to keep more than a minimum number of parts available for "withdrawal."

The second principle is: ***Components should be available when required.*** When a JIT workcell withdraws parts from another workcell, it leaves behind a *"production ordering Kanban."* This type of Kanban orders the manufacture of parts to replace those taken.

Parts must be available when required, but maintaining large stocks of those parts is undesirable, so the workcell that manufactures the parts must respond to production ordering Kanbans as quickly as possible. Employees are encouraged to identify and eliminate unnecessary operations or unnecessarily complex operations.

The above two JIT principles require a different approach to production from that traditionally applied in North America. *Workers and equipment at workcells must be multi-functional.* Each JIT workcell can produce several different types of (related) parts.

---

[1]Monden, Yashuhiro, *Toyota Production System,* Industrial Engineering and Management Press, Institute of Industrial Engineers, Norcross, Ga., 1983.

This allows workcells to respond to changing production requirements while (for example) the need for one of its products grows and for another product decreases.

For workcells to be multi-functional, each worker must be encouraged to develop as many skills as possible. Workers can then change production duties when the constantly changing production mix requires them to. Expansion of workers' skills extends to their being trained to work at other workcells. Having worked at several different steps in production, employees can offer more valuable advice on how to increase production efficiency.

Equipment, too, must be "multi-functional," and quickly changeable so that it can produce different types of parts on short notice. Reduction in the number and complexity of machine-setup operations is emphasized. Employees are encouraged to identify and implement changes that save setup time. As setup time is reduced, it becomes possible to change machine setup more frequently. Frequent setup changes mean more immediate response to newly-received production ordering Kanbans. It also means that it is economical to manufacture fewer parts per setup. Production ordering Kanbans can then specify fewer parts, and fewer unneeded parts are manufactured.

The third principle is: *Unnecessary costs should be eliminated.* While all manufacturers try to do this, Toyota's JIT application is unique. We have already noted that production employees are encouraged to suggest and implement production simplification and speed-enhancing changes.

Toyota employees are also expected to participate in *quality circles (QC)*. Quality circle teams consist of representatives from several different parts of the company. A single quality circle may include members from the assembly floor, the machine shop, marketing, and top management. Decisions made in a quality circle must be unanimous, but nobody outranks anybody else while in the quality circle! Quality circles examine any aspect of the company that they select to examine, call in expertise they see themselves as lacking, and issue orders (not recommendations) to the company. A worker on an assembly line can suggest a production system change to his/her quality circle and (if the QC agrees) can see the change carried out. Radical as it sounds, it works because workers soon recognize that their own jobs depend on how well they run the company.

The fourth principle is: *Product quality should be improved.* This means the reduction of defects to zero. It results in improvement in value of the product to the consumer.

Toyota's JIT philosophy joins the goals of cost reduction and quality improvement. Quality circles, as the name suggests, are expected to improve product quality, not just reduce costs.

At Toyota, employees are encouraged to take "ownership" of the manufacturing process while on the production floor as well as in the quality circle. They are *allowed to stop production* when they detect defects, and in fact, it is expected that they *will stop production* occasionally for this purpose. It is assumed that there will always be new reasons for faulty products, and if an employee doesn't find defects, then that employee isn't looking!

As soon as production is halted, the employees where the defective parts were produced are expected to solve the problem, calling in supervisors and experts only as necessary. Remember that this is in a JIT production system. A production delay of only a few minutes can bring the whole plant to a halt! Halting production is actually seen as desirable, because it reinforces the idea that even one defect is too many to be tolerated.

JIT methods, which give control of the company to all employees, work. JIT methods work because they *encourage use of all the brains in an organization,* including those of the workers who work day-in and day-out with the product and with the production equipment.

SPC for Simplification

Sometimes it is desirable to use **Statistical Process Control (SPC)** techniques to monitor production quality. SPC quality monitoring can detect that it is probable that a defect will be produced, *before* it is produced.

Parts supplier conformance to ISO 9000 standards is becoming a requirement in many companies' purchasing policies. The ISO's (International Standards Organization's) 9000 series of standards specify SPC quality control **procedures.** When a purchaser buys parts from a vendor who meets these standards, the purchaser knows that the parts meet a high standard of quality. There are service organizations that will examine a company's SPC equipment, training, and procedures, and provide certification that the company meets ISO 9000 standards.

SPC techniques require monitoring and recording of part dimensions and other measures of quality while parts are being produced. Measurements are plotted on graphs that show whether the manufacturing process is **under control,** or is tending toward producing defective parts, or is becoming erratic.

SPC graphs can be used to identify manufacturing processes that need improvement. *Solving production problems makes it simpler to produce quality parts.*

10.2.2 Automation

Eventually, processes where **automation** is desirable will become evident. Some noncomplex processes should be automated first, even if the benefits of automation aren't the highest at these processes. During automation of simple processes, employees will learn improved automation skills that they can then apply to bigger projects.

Many experts advise designing all automation such that it can be integrated later into a fully Computer Integrated Manufacturing system. The unfortunate fact is that, during the automation of the first few processes, employees probably won't understand automation well enough to plan for computer integrated manufacturing. Retrofitting these early automated processes for integration later is probably cheaper than designing-in inferior integration capabilities now.

Having learned the material in this text, and having been involved in the process simplification task, the reader is ready to select components and controllers from industrial suppliers and to build simple workcells. With experience and a growing familiarity with proprietary components, the automated system designer will find that more complex automated system design is possible.

10.2.3 Integration

After building one or two independent "islands of automation," it becomes reasonable to think about how these and future islands can be **integrated to work together.**

The manufacturing system (consisting of both automated and manual processes) must be examined as if it consists of discrete functional blocks. Essential information flow between each functional block must be identified. A computer system (hardware and software) will then be specified to handle the information flow. The components of the computer-integration system can be acquired and installed one piece at a time, as they are required and as personnel become experienced at selection and installation.

10.3 SAFETY CONSIDERATIONS

People are an inventive lot. Some invent ways of making equipment safe to use, and others invent ways of circumventing the safety features.

Designing safety into an automated system can be difficult. Few standards exist. Government and corporate standards that do exist sometimes result in reduced safety. There are no rules that always work. Each installation poses its own set of problems.

The best safety feature is one that *doesn't need to be circumvented*. If the "safety feature" makes it difficult for someone to do his or her job, that worker may find a way to disable the safety feature, leaving no protection for other workers.

One almost universally-accepted safety rule is that machinery capable of causing physical damage should be **"safety interlocked."** A safety interlock could be installed on a gate separating an operator from a dangerous piece of equipment. If the gate opens, the interlock disconnects the equipment's main power supply. The guard fence around a robot can, for example, include a switch that cuts robot power when the gate is opened. The decision to include a safety interlock forces the designer to consider the consequences of having it in place. How will the robotic maintenance worker, who sometimes needs to observe the robot closely, handle the safety interlock? If the switch must be disabled for maintenance, how can it be *ensured* that the safety feature is reenabled when the maintenance is finished?

Safety interlocks and emergency stop buttons lead to other difficult decisions. If the equipment grips and moves materials, should the materials be released when power goes off? What happens if the material drops? When power goes off, material holders may inherently release materials. Should **"fail-safe brakes"** be installed in the material holder to prevent materials from being dropped?

A fail-safe brake is a normally-actuated brake. The brake only releases, allowing motion, when power is on. Fail-safe brakes can be attached to conveyor motors, or to any other rotating shaft, in the material-handling system. Hydraulic systems can also be connected as "fail-safe" systems. A valve that needs to be powered open is installed in an actuator's return line.

Pneumatic systems can be fitted with two- or three-way valves in their air supply line as part of a safety interlock. When power goes off, the valve returns to its normally-closed position, cutting off air to the system. The designer may select a three-way valve so that the *system pressure exhausts* when the valve closes.

The problem with using a two-way valve, which traps air pressure in a pneumatic system when it is off, is that moves that were started before power was cut will continue if a detent valve has been actuated. Other actuators, if controlled by *actuated* spring-return valves, will suddenly start to move when power goes off. If pressure is retained (or if it exhausts slowly from a pneumatic system when power goes off), the designer should carefully select valve types so that actuators return or retain their positions, whichever is safest.

Safety features must be carefully thought out for each application, but the *safety features should be as uniform as possible among all machines.* The shape and location of the emergency stop button, for example, should be the same at all workcells. Sometimes the need for commonality conflicts with unique needs at a single workcell. Can an acceptable compromise be found in redesigning the workcell?

Education is still the most important component in any safety system. The user should be able to recognize when the equipment is behaving in an unsafe manner, and should recognize why some methods of using the equipment are unsafe. The user of the system, and the user's supervisors, should have a healthy respect for the danger and should be willing to sacrifice production rather than safety.

COST JUSTIFICATION FOR AUTOMATION

Traditionally, equipment isn't purchased for production purposes until the corporate accountants are convinced that buying the new equipment will result in savings in production costs.

Savings are often expressed in terms of how long it will take for the new equipment to pay for itself. This "payback period" must be shorter than the expected life cycle of the product. Some products (this year's fashion in dresses, the latest personal computer) have shorter life cycles than others (blue jeans, a basic telephone).

Well-understood formulas exist to calculate such savings, and they work well when you are comparing (say) a drill press against a manual drilling operation, or a fast milling machine against a slower one. When trying to justify a completely new manufacturing system against the old methods, the economic justification techniques aren't as easy to use. The standard techniques require more difficult-to-acquire numerical estimates, and often don't account for the value of changes in product quality.

To illustrate, one company needed to prove that a proposed automated system would pay for itself in five years. It took two years to prepare the proof to the satisfaction of the reluctant accountants. Once installed, the system paid for itself in just over one year, half the time it took to convince the accountants! Part of the reason for the fast payback period was due to improvements in product quality and the sales increases that resulted.

The following formula can be used to find a proposed system's payback period, if all the costs and benefits can be determined.[2] The formula also can be rearranged to

---

[2]*Tool Design*, 3d ed. Donaldson, LeCain, and Goold. McGraw Hill, 1973, pp. 574–576 (Originally from: A Treatise on Milling Machines, 3d ed, Cincinnati Milacron Inc., Cincinnati, OH 1951.)

calculate how much can be paid for the new system, the required production rate of the new system, or the savings per part if the new system is implemented.

$$a = \frac{C}{Ns(1+L) - Si - C(i+u+t)}$$

where:

a = payback period (years)
C = cost of the proposed system ($)
S = unrecoverable cost of scrapping the old system ($)
N = number of parts/year to be produced by the new system
s = savings per part by changing to the new system ($)
L = percent overhead charged against parts (decimal %)
i = annual cost of money (decimal %)
u = annual upkeep cost (decimal % of C)
t = tax, insurance, etc., costs (decimal % of C)

EXAMPLE 1

A proposed automated workcell will cost $2,500,000 (C = $2,500,000).

The old manual manufacturing equipment is now worth $750,000 according to the company accountant. We can get only $50,000 for it, by reusing some parts (instead of buying new) and by selling other parts as used equipment or as scrap metal. The unrecoverable cost (S) is $700,000.

Working at full speed, the new workcell should be able to process a part every 10 seconds. We estimate that work stoppages for an assortment of reasons will waste 20% of the available time per day, so it will take an average of 12 seconds to process a part. At one part every 12 seconds, the new workcell can produce 5 parts per minute (300 parts per hour, 2400 parts per 8-hour shift, or 600,000 parts in 250 single-shift days per year). Marketing says they can sell 550,000 parts per year, which is why this workcell was designed as it is. Assume we will make as many parts as we can sell. N = 550,000.

The automated workcell will allow the parts to be produced with lower labor costs, and there will be less waste of materials. The savings per part (s) is $2.25.

The department where the processing of these parts takes place contributes to the heating and lighting costs of the plant. The floor space has an annual value too. These "overhead" costs have been calculated as a percentage of the production costs in each area of the plant. In this department, overhead is estimated at 35% (L = 0.35).

The accountants tell us that if we borrow money to finance the workcell, interest charges will be 9%, so i = 0.09.

Maintenance costs for the workcell are estimated at $500,000 per year. This includes all lubrication, replacement parts, and maintenance labor. This annual upkeep cost is 20% of the purchase price (C), so u = 0.20.

Insurance and other annual costs of owning the automated workcell will work out to 6% of its value per year, so t = 0.06.

The payback period works out as follows:

$$a = \frac{C}{N_s(1+L) - S_i - C(i+u+t)}$$

$$= \frac{2{,}500{,}000}{(550{,}000 * 2.25 * (1 + 0.35)) - (700{,}000 * .09) - (2{,}500{,}000 * (.09 + .2 + .06))}$$

$$= 3.41 \text{ years.}$$

So for 3.41 years, the savings of $2.25 per part will go toward paying for the equipment. After 3.41 years, these savings will be increased profit for the company.

Often, the above formula is restructured to use labor and material costs and a fixed allowable payback period to find "s" (the savings per part by changing to the new system). More recently, quality control advocates have shown that changes in product *quality* can lead to savings, too.[3] The following costs are associated with lack of quality in a product, and can be reduced if product quality improves.

1. **Costs due to internal failure.** The costs to the company associated with the defective products that are caught before they leave the plant. They include the costs incurred in scrapping or reworking defective parts, the cost of researching reasons for defects, and the cost of correcting the source of defects.

2. **Costs due to external failure.** The costs associated with defective parts that make it to the customer. These costs include the costs of paying out on warranties or providing replacements to the customer, and the cost of investigating customer complaints, including the cost of corresponding with the customer. They also include the significant cost of lost sales and of lowering retail prices to try to keep disgruntled customers.

Because many customers don't complain but simply buy elsewhere once we sell them a defective product, these costs can be difficult to quantify.

3. **Appraisal costs.** The costs of doing inspection, assuming that defects will always exist and will need to be "caught." These costs include the inspection labor costs for incoming, in-process, and final inspection, and also the cost of tools and shop floor space required by inspectors.

4. **Prevention costs.** These are costs incurred to prevent defects from being produced and being shipped to the customer.

These costs can include the training of inspectors, operators, and suppliers to avoid defects; the costs of developing plans to handle defects; and the costs incurred in determining the probable quality problems inherent in new products.

---

[3] J.M. Juran and F.M. Gryna, *Quality Planning and Analysis,* 3rd ed., McGraw Hill, New York, NY 1993.

Some estimates have placed the cost of non-quality as high as 20% of the price that a consumer pays for a product!

Once the above values have been calculated (or estimated) for a new production system's effect on quality, then they can be included in the costing formula.

EXAMPLE 2

Marketing estimates that if we can eliminate failures of our products in the field, we will increase sales by 5%. This means that we must produce 575,000 parts per year. "N" changes to 575,000.

The proposed system already has a capacity of 600,000 parts per year, so no design changes are required to increase capacity. Improving product quality, however, means we have to pay for a better-quality production system. The cost of the proposed system (C) increases from $2,500,000 to $2,750,000.

We find that the improvement in consistent quality will reduce quality control costs, warranty and field replacement costs, and the cost of tracking down sources of defects. We save an additional $0.30 per part. Increased overtime payments (because of increased production quantities) will reduce these savings per part by $0.05. "s" increases by $0.25 to $2.50.

Higher quality replacement parts, and increased use of the workcell increases its annual upkeep costs by $33,000 per year to $533,000. This is 21.2% of the system's (new) purchase price, so "u" changes to 0.212.

Using these revised values in the formula previously used to calculate the payback period for the proposed workcell yields:

$$a = \frac{2{,}750{,}000}{(575{,}000 * 2.50 * (1 + .35)) - (700{,}000 * .09) - (2{,}750{,}000 * (.09 + .212 + .06))}$$
$$= 3.11 \text{ years}.$$

So by building a system to produce higher quality parts, the company will start to see increased profits from the new workcell in only 3.11 years, not the 3.41 years originally calculated. At $2.50 per part and at 575,000 parts per year, over the extra 0.3 years, additional profits work out to be:

$$575{,}000 * \$2.50 * (3.41 - 3.11)$$
$$= \$ 431{,}250$$

There is yet another adjustment to be made to the costing formula as it applies to automation. In the Social Effects of Automation section of this chapter, we will discuss the role that industrial automation is playing in unemployment, and what effect unemployment has on a manufacturer's profits. The effects are difficult to calculate and "plug into" the costing equation. We will see, however, that implementation of automation with resultant unemployment, and the values used in the costing equation (i.e., production quantities, savings per part, overhead cost, interest rates, upkeep costs, and tax rates) are unavoidably related.

SOCIAL EFFECTS OF AUTOMATION

This section includes some personal opinion. Some topics are controversial. We include this section because the computer revolution, including industrial automation, often seems to be proceeding without our legislators' knowledge or participation.

It has been said that a good automation or CIM implementation can make work better for employees. Humans may no longer have to do dirty or dangerous jobs. Employees are freed to use their minds instead of their bodies. The self-image of employees in automated workplaces rises as their skills improve with training. This is all true.

"If we don't automate, there won't be any jobs!" Often this, too, is true. Successful automation can increase local employment while providing higher caliber jobs, due to increased sales of the higher quality products.

The Problem

Automation often increases production while **reducing the labor** needed in production. Computer Integrated Manufacturing can lead to reduction in even the numbers of *non-production* employees needed. Managers, engineers, technicians, and office staff can join production workers in the "no longer required" category.

Automation allows fewer people to produce more goods. How might industry respond to the changes in productivity? They have four options.

☑ One option is to *find new markets for goods, so that we can **increase** the size of the workforce in manufacturing.* Market control is only shifted when one company takes market share away from another. Increasing the size of the *global* workforce requires increasing the global market size. This requires increasing the whole world's wealth so that the average person can afford more. So far, no organization has been successful in even slowing down the rate of starvation except in tiny sections of the world. It seems a little unrealistic to ask industry to increase the world's wealth.

☑ A second option is to *find only enough new market share to **maintain** the current levels of employment.* Since goods can be produced and distributed cheaper now, consumers should be able to buy more without being wealthier. In effect, increased profits from lower production costs would all have to be passed on to customers. Asking industry to give away *all* its profits to the consumer seems unfair.

☑ A third option is to *use increased profits from reduced production costs to **maintain** the current level of employment, even though unchanged production levels mean that all employees are no longer needed.* In effect, the increased profits must all be transferred to the workforce. Asking industry to give all its increased profit to the workforce is as unfair as asking them to give it all to their customers. Many people try to use variations on this approach, however. Some Japanese corporations promise that employees will never be laid off. While Japan's exports continue to grow, this hasn't yet required that *all* new profits be used to keep redundant employees.

Government intervention in society's economic matters is sometimes intended to transfer increased corporate and personal profit to displaced workers. Unemployment insurance payments, legislated severance pay, and (eventually) welfare payments are examples. These efforts never replace all the displaced employee's income, sometimes don't take all the increased profits from industry, and always waste money by (for example) requiring new government employees to administer the transfer.

Distributing increased profits to unneeded employees (or to customers) could work if there was enough *new* wealth so that manufacturers could keep some profits without reducing the amount distributed to consumers and employees. We have already dismissed increasing the world's wealth as wishful thinking.

Some countries try to borrow the extra wealth. Many third world countries, and even some developed countries (New Zealand, for example), have discovered that eventually the loans and the interest have to be repaid. Escalating interest payments must be passed on to citizens through tax increases and/or service cuts. Governments benefit from inflation if they owe money, because the money they owe becomes worthless, so they may actually allow inflation and interest rates to remain high.

A fourth option is to **reduce** *the size of production workforces.* During the 1980s and early 1990s, the percentage of North Americans employed in manufacturing fell. While some of the fall was due to loss of market-share to manufacturers in other countries, some of it was due to manufacturers laying off workers displaced by automation and computerized processes.

Eliminating a worker doesn't mean eliminating a person, so society's costs go up as automation brings manufacturing costs down. In fact, industry is negatively affected by this method of increasing profits. What happens when the total number of production workers drops?

The displaced worker *may find another job in manufacturing.* But if there are fewer production jobs, this only displaces another worker. Retraining may be required for some individuals to return to work. The benefits of increased production capabilities are reduced by the cost of training (including training for jobs that don't exist).

Some displaced production workers *find employment in service industries.* Service industries often consume the community's labor pool without providing new wealth. Services are unlikely to be exported. Service industries rarely produce a product that can be described as increasing a community's (or the world's) wealth.

A cooked hamburger has more value to a hungry person than an uncooked one. When viewed this way, some jobs usually viewed as service industry jobs take on the aspect of production jobs. As these **service/production industries** grow, open branch plants (franchises), export raw materials, and bring profits back to their home communities, they *do increase the community's wealth.*

Some social service industries increase a community's overall wealth by *improving the quality of life.* More often, social workers are needed to just slow a decrease in quality of life, using up labor capacity that could otherwise be producing wealth.

Some service industries only consume wealth. Supermarket food is often wrapped in as many as seven separate disposable wrappings while it is carried home. The labor and materials to provide this packaging provide no benefit for the consumer.

Service workers are paid significantly less than production workers. Advancement in a service industry is usually slower than in production jobs. *The average service employee cannot afford to buy as much as a production worker.*

Some displaced workers will *remain unemployed.* Most governments have some form of social assistance program that taxes those with money to partially replace the income of those who have lost their jobs. Supporting the unemployed costs manufacturers money and takes money away from consumers who might otherwise buy the manufacturer's products. When the income of displaced production workers drops, they cannot afford to purchase as much. As the percentage of the population in the ex-production worker category increases, the population buys less. *Employers sell less.*

What Can We Do to Change Things?

From the discussion so far, it would appear that automation can only make society increasingly poor. We have no intention of abandoning the reader with that thought. We only hope that the discussion so far has convinced the reader that computers and automation are changing society as drastically as society was changed during the industrial revolution. The change may be even more drastic because, this time, less of the world will remain untouched.

During the industrial revolution, society found it needed legislation to correct problems such as child labor, unsafe working conditions, abuse by industry owners, overcrowding in cities, and other symptoms of the revolution. During our computer revolution, we have unemployment, under-employment, abuse by the owners of data, failure to transfer technological knowledge to large groups of people, and other problems. Everybody must be prepared for the significant corrections that society needs.

If human productivity is increasing, all humans SHOULD be seeing more benefits! Instead, automation and computer control has been reducing the actual amount of production. Instead of there being more benefits, there are fewer. How can society's approach to automation be modified to realize the potential additional benefits?

Let's take another look at the four ways this chapter said we could react to increased automation:

1. Increase markets, and thus production and employment, by increasing the world's wealth.

2. Sell goods at lower prices, increasing demand as production increases so that employment stays the same.

3. Use increased profits from unchanged production levels to maintain workers rather than laying them off.

4. Lay workers off and keep the profits.

During the industrial revolution, some societies used a fifth option, violence, to reverse the concentration of wealth into fewer hands, especially when there wasn't enough for the poor to live on.

To avoid economic problems that might lead to violent solutions, we need practical methods to harness our increased productivity so that we can all share in the benefits. We won't discuss just maintaining current production levels or profit levels (options 2 and 3 above). Increased productivity should allow us to do better. Option 4 (layoffs) is destructive.

We are left with "increasing the world's wealth," which was so easy to dismiss when first mentioned. The industrial revolution, when properly harnessed, led to a wealthier world and the computer revolution can do so, too. We just have to recognize and harness the changes that are occurring.

As an example, a very real impediment to increasing world wealth is that some very important natural resources are being lost (not necessarily due to the computer revolution). Fertile farmland, clean water, living space, oxygen, and the ozone layer are some examples of basic raw materials that are disappearing. The conservation of natural materials requires planetwide projects.

One suggestion for harnessing this change starts by recognizing that computers and communication networks have created systems that allow an easy exchange of ideas and the availability of large amounts of data. Industrial automation has made a lot of intelligent and highly-trained individuals available ("unemployed" if you prefer). These individuals could be employed on projects to conserve natural materials, instead of working in the current crop of service industry and government clerical make-work jobs. Conservation projects deserve the attention of the best minds, so there is no reason *good* service industry jobs can't be created in conservation projects. The pay for these jobs must be high enough to compete with industry for the best people.

We, in industry, should be encouraging government to react to the computer revolution and we should be telling government how to act, even if it costs us something. We should be doing this in part because we live on this planet, but also because we want to ensure that industry continues to have a market to sell to.

Industrialists are familiar with the idea of investing money for future expansion. That's how they get wealthy. In today's world market, *national* market development isn't sufficient. Industry has to invest in the *world's* future.

*It would be a shame if we automated, emigrated, legislated, **AND** evaporated!*

10.6 SUMMARY

An organization that understands the technologies available and recognizes that they can be used to improve quality and thus reduce cost is prepared to begin the automation process.

Before automating anything, a manufacturer should attempt to involve all employees in process **simplification.** Toyota's **Just in Time** system and **Statistical Process Control** are two examples of tools that might be used.

Only after manufacturing processes are understood, simplified, and controlled can **automation projects** be begun. After some experience has been gained in automation, **computer integration** of the automated processes should be undertaken.

Equipment should *not only be safe, it should be foolproof.* There are no safety features that always work, so designers should be careful to consider what safety features will be needed, what effect they will have on day-to-day operation and maintenance of the equipment, and how the system will react when safety features are activated. The use of **safety interlocks** and **fail-safe brakes** should be considered.

Well-understood **formulas for evaluating costs and benefits** of new equipment exist. They can be adapted to evaluate proposed automated systems. **Costs and benefits of quality improvements** should be included. Some societal effects of industrial automation should be involved in the calculation but may be difficult to estimate.

Automation allows more higher-quality products to be produced at lower cost, so **society should benefit.** The computer revolution, including industrial automation, is causing social changes, including **unemployment and reduced income,** for large segments of society. All of our society is paying the cost, including manufacturers, so they aren't seeing the benefits they expected from automation. *Society needs to find productive, wealth-generating work* for the human resources no longer required in manufacturing.

REVIEW QUESTIONS

1. When should an organization automate?
2. How can an organization ensure that its employees are ready for an automation project?
3. Automation is only the third step in a three step process.
   a) What step must be begun well before automation projects are undertaken?
   b) What is the step after simple automation and when should it be seriously begun?
4. What are the four principles embodied in Toyota's JIT philosophy?
5. What does "multi-functional" mean as it applies to workers in Toyota's JIT philosophy?
6. Why are "multi-functional" workers more valuable?
7. What is commonly plotted in an SPC system? What is watched for in the output plot?
8. If the corporate accountants decided that the manufacturing system in the example on pages 281–282 had to be paid for after only 3 years, not 3.41 years:
   a) Show how you would rewrite the equation to calculate cost of the system (C) based on the other factors.
   b) Calculate the revised allowable cost of the proposed system for it to be paid back in 3 years, if all other variables remain the same.
   c) A designer of an automated manufacturing system for a well-understood manufacturing process should be able to reduce the manufacturing cost, within reason, without reducing product quality. If, however, the cheaper production system resulting from your calculations in part b) resulted in a loss of quality so that sales went down by 1%, how long would the system payback period be extended?
9. You have worked out that you can save $1.25 per final product by improving the final quality of the product. The estimate includes production quantities of 25,000 per year. The system that you now use to manufacture the product has a value of $50,000, of which you can salvage $10,000. Overhead costs would be 25% in the department where the new manufacturing system would be installed. The cost of the money needed to build the new manufacturing system is 6%. If upkeep of the system can be held to 20% of the system cost, and insurance and other costs held to 8%, and if the new system must pay for itself after no more than 3 years, what is the maximum amount you can invest in a new manufacturing system?

INDEX

Absolute optical encoders, 99–100
AC generators, 100–101, 102
AC motors
    for control applications, 58–68
    control for, 65–68
    induction motors, 52, 60–64, 237
    single-phase, 61–62
    stopping of, 68
    synchronous, 50, 64
Acceleration sensors, 101–102, 103
Accelerometers, 101–102, 103
Accumulator, 199
Actuators
    air-control valves and, 31–33, 34
    air-power, 27–31
    air supplies and, 33–34, 35
    as automation components, 8, 11–12
    conveyors and, 36, 37, 38
    feeders and, 37–38, 39
    heat, 21
    hydraulic, 35–36, 238–239
    indexing tables and, 37, 38–39
    introduced, 21
    light, 21–25
    piezoelectric force generators, 25
    pneumatic, 237–238
    power transmission components and, 39–40, 41
    rotary, 30
    selective compliance articulated robot actuator (SCARA), 233–235
    solenoids, 25–27
    special-purpose, 41–44
A/D converters (analog to digital converters or ADC), 122–123, 208–212
Adaptive controllers, 157
Adaptive direct synthesis control (ADSC), 183
ADC (analog to digital converters or A/D converters), 122–123, 208–212
Address bus, 198
ADSC (adaptive direct synthesis control), 183
AIM USD-2 (Automatic Identification Manufacturers Uniform Symbol description 2) code, 105

Air-control valves and actuators, 31–33, 34
Air cylinders, 27–31
Air motors, 31
Air-power actuators, 27–31
Air supplies and actuators, 33–34, 35
Aliasing errors, 121–122
Alternating signal sensors, 102–104
Amplifiers
    derivative, 223, 224
    error, 152, 223, 224
    operational, 57, 100, 101, 207, 222–224. *See also* Controller algorithms
Analog (continuous), defined, 204
Analog computers, 225
Analog controllers, 9, 225–226
Analog I/O, 208–212
Analog to digital converters (ADC or A/D converters), 122–123, 208–212
Angular grippers, 241
Application programs, 203–204, 246
Application service layers (OSI model), 217–218
Application specific integrated circuits (ASIC), 199
Applications layer (OSI model), 217–218
Architecture of digital computers, 13–14, 16, 197–199
Arm of robots, 240–241
Armature voltage control, 56
Armatures in motors, 48, 50
Artificial intelligence programming languages, 253
ASIC (application specific integrated circuits), 199
AS/RS (automated storage and retrieval systems), 43–44, 271
Asynchronous communications, 214
Automated gauging sensor systems, 104
Automated storage and retrieval systems (AS/RS), 43–44, 271
Automated workcells, 256–264, 265
Automatic Identification Manufacturers Uniform Symbol description 2 (AIM USD-2) code, 105
Automation
    benefits of, 3, 274–275
    components in, 8–12

289

Automation, cont.

    control of, 6–8
    cost justification for, 280–283
    employment and, 284–287
    environment for, 3–6
    hard, 7
    of individual processes, 4
    interfacing and. *See* Interfacing
    labor force and, 284–287
    organizational approach to, 275–279
    overview, 4–6
    safety considerations, 279–280
    signal conditioning and, 12–16, 204, 205
    social effects of, 284–287
    soft, 7–8
    unemployment and, 284–287
    when to automate, 274–275
    workforce and, 284–287
Automation control, defined, 7

Backlit surfaces, 114, 125–126
Ballnuts, 40, 41
Ballscrews, 40, 41
Band heaters, 21, 22
Bar code readers, 104–105, 266
Bar codes, 104–105
Baseband networks, 215
Basic input/output system (BIOS), 201–202
Bearings in motors, 48
Belts, timing, 40
Biasing errors, 121–122
Bin picking problem, 140–141
Binary numbering system, 99–100
Binary outline images, 125
BIOS (basic input/output system), 201–202
Bipolar transistors, 221–222
Blooming of images, 130
Blowers, 34
Board cleaning machines, 43
Bode diagrams, 103, 158–161
Bowl (rotary vibratory) feeders, 38
Brakes, fail-safe, 279
Braking motors, 57, 68
Brightness distortions, 130
Broadband networks, 215
Brushless DC motors, 70–73, 237
Bubbler type sensors, 88

Bus networks, 216
Buses
    address, 198
    control, 198
    data, 197

CAD (computer aided design), 270
CAM (computer aided manufacturing), 42
Cameras, 113, 116–120
Capacitance of control systems, 169, 170
Capacitive position sensors, 93, 94
Capacitive proximity sensors, 83, 84
Capacitor run motors, 64
Capacitor start motors, 63–64
CAPP (computer-aided production planners), 270
Carrier sense multiple access with collision detection (CSMA/CD) standard, 216
Carrierband networks, 215
Cartridge heaters, 21, 22
Cascade controllers, 184–185
Cathode ray tube (CRT) displays, 24
CCD (charge coupled device) cameras, 118–119
Cell controllers, 196, 266
CEMF (counter electromotive force), 51, 52
Central controllers, 196–197
Central processing unit (CPU), 197, 199–200, 244–245
Chain conveyors, 36, 38
Charge coupled device (CCD) cameras, 118–119
CIM. *See* Computer integrated manufacturing (CIM)
Circuit and component testers, 43
Circular splines, 40
Closed loop control systems, 6, 7, 151–155, 156
CMM (coordinate measuring machines), 104
CNC (computer numerical control) equipment, 42
Code 39 standard, 105
Communication ports in controllers, 10
Communications and networks, 214–218, 268
Communications programs, 217–218, 248
Commutators in motors, 48
Comparators, 224
Compiled programs, 204, 252–253
Component insertion machines, 42
Component soldering machines, 42–43
Compound wound motors, 55
Compressing images, 128–129
Computer aided design (CAD), 270
Computer aided manufacturing (CAM), 42

Computer-aided production planners (CAPP), 270
Computer integrated manufacturing (CIM)
   automated storage and retrieval systems (AS/RS) in, 44
   defined, 257
   environment for, 4–6
   example, 268–271
   islands of automation and, 264–265
   overview, 268
   real world and, 271
   sensors in, 104
Computer numerical control (CNC) equipment, 42
Computers
   analog, 225
   digital. *See* Digital computers
Condition code register, 199–200
Configure, defined, 204–205
Connectedness, 133, 137
Constraint control, 185
Contact bounce in switches, 80–81
Continuous (analog), defined, 204
Continuous path robots, 251–252
Control bus, 198
Control levels, 193–197
Control systems
   capacitance of, 169, 170
   closed loop, 6, 7, 151–155, 156
   deadband of, 167–168
   deadtime of, 169, 170
   distributed, 10, 248
   feedforward, 184–185
   hysteresis in, 168
   open loop, 6, 7, 151, 156
   predictive, 184–185
   resistance of, 169
   saturation in, 168
Controlled path robots, 251
Controller algorithms
   on/off control, 170, 171
   proportional (PE) control, 170–173
   proportional derivative (PD) control, 175–177
   proportional integral (PI) control, 173–175
   proportional integral derivative (PID) control, 178–181
Controller configurations
   adaptive direct synthesis control (ADSC), 183
   direct synthesis control (DSC), 183, 184
   feedforward control systems, 184–185
   nested control loops, 183

   predictive control systems, 184–185
Controllers
   adaptive, 157
   analog, 9, 225–226
   as automation components, 8–10
   cascade, 184–185
   cell, 196, 266
   central, 196–197
   communication ports in, 10
   and controlled systems, 181–182
   differential gap, 170, 261, 263
   digital, 179–180, 186–189
   fuzzy logic, 181, 187
   integral, 175
   interfacing
      to controllers, 14–15
      to other components, 15–16
   motor, 48–50
   predictive, 175–177
   ratio, 185
   reset, 173–175
   in robots, 244–248
   silicon-based, 221–225
   workcell, 194–196, 258–264, 265
Controlling DC power levels, 57–58
Conveyors, 36, 37, 38
Coordinate measuring machines (CMM), 104
Coordinated point to point robots, 251
Co-processors, 199
Cost justification for automation, 280–283
Counter electromotive force (CEMF), 51, 52
CPU (central processing unit), 197, 199–200, 244–245
CRT (cathode ray tube) displays, 24
CSMA/CD (carrier sense multiple access with collision detection) standard, 216
Cutters, 41
Cycloconverters, 67, 68

D/A converters (digital to analog converters or DAC), 208
Data bus, 197
Data link layer (OSI model), 218
Database management systems (DBMS), 196, 270
Databases, 196
dB (decibel) unit, 160
DBMS (database management systems), 196, 270
DC motors
   brushless, 70–73, 237

Index   **291**

DC motors, cont.
    for control applications, 54–58, 59
    introduced, 50–51
    permanent magnet, 54, 237
    silicon controlled rectifier (SCR) motors, 70, 71
    speed control for, 55–57
    stopping of, 57
    wound field, 54–55
DC power level control, 57–58
Deadband of control systems, 167–168
Deadtime of control systems, 169, 170
Debouncing switches, 81
Decibel (dB) unit, 160
Decision support software (DSS) programs, 269–270
Degrees of freedom, 233–236
Derivative amplifiers, 223, 224
Derivative time, 177
Detent switches, 79
Detent valves, 32–33
Differential gap controllers, 170, 261, 263
Differential pressure sensors, 86, 89
Digital (discrete), defined, 204
Digital computers
    architecture, 13–14, 16, 197–199
    central processing unit (CPU), 197, 199–200, 244–245
    I/O. *See* Input/output (I/O) devices
    memory. *See* Memory
    open architecture of, 13–14, 16
    operating systems, software, and memory organization, 200–204
    personal computers (PC), 13–14, 199
    PLC. *See* Programmable logic controllers
    sizes, 199
Digital controllers, 179–180, 186–189
Digital level, 204, 205, 207
Digital to analog converters (DAC or D/A converters), 208
Dimensional distortions, 130
Diodes, 221, 222
Direct drive robots, 237
Direct drive servosystems, 72
Direct numerical control (DNC) equipment, 42
Direct synthesis control (DSC), 183, 184
Discrete (digital), defined, 204
Dispensers, 31

Displays
    cathode ray tube (CRT), 24
    gas plasma, 24
    light emitting diode (LED), 21–23
    liquid crystal (LCD), 23–24
Distributed control systems, 10, 248
Disturbances, 163, 185
Diverters, 26
DNC (direct numerical control) equipment, 42
Doppler effect speed sensors, 101, 102
Dot-matrix printers, 26
Double pole (DP) switches, 80
Double throw (DT) switches, 79
DP (double pole) switches, 80
DSC (direct synthesis control), 183, 184
DSS (decision support software) programs, 269–270
DT (double throw) switches, 79
Dual purpose end effectors, 241
Dual ramp analog to digital converters, 209–210, 211
Duty cycle of solenoids, 26–27
Dynamic braking, 57, 68

EAPROM or EEPROM (electrically erasable programmable read only memory), 199
Economical analog to digital converters, 210, 212
EEPROM or EAPROM (electrically erasable programmable read only memory), 199
Electrically erasable programmable read only memory (EAPROM or EEPROM), 199
Electrodes, 23, 24
Electromagnetic end effectors, 241
Electromagnets, 25
Electronically commutated motors, 68–73
Employment and automation, 284–287
End effectors in robots, 241–242
Erasable programmable read only memory (EPROM), 198–199
Error amplifiers, 152, 223, 224
Error signals, 152, 163–165
Errors
    aliasing, 121–122
    biasing, 121–122
    following, 165
    quantization, 122–123, 124
    residual, 172
    sampling, 121–122
    in sensors, 76

Errors, cont.
    steady state, 158, 172
    type 1 and type 2, 131, 143
    unit error signals, 163–165
Excited coil synchronous motors, 64

Fail-safe brakes, 279
FDDI (fiber digital data integration) standard, 216
Feedback signals, 152, 156
Feeders, 37–38, 39
Feedforward control systems, 184–185
FET (field effect transistors), 222, 223
Fiber digital data integration (FDDI) standard, 216
Fiber optic networks, 216
Field effect transistors (FET), 222, 223
FIF (fractal image format), 128
Filter-regulator-lubricators (FRL), 34, 35, 237, 244
Filters
    gaussian, 131
    polarizing, 23, 24
Flash analog to digital converters, 210, 211
Flash memory, 199
Flex splines, 40
Flexible manufacturing systems (FMS), 104, 257, 266–268
Floating point control, 170
Flow control valves, 244
Flow transducers, 88–89, 90
FMS (flexible manufacturing systems), 104, 257, 266–268
Following errors, 165
Footers of messages, 214
Force transducers, 86–88, 89
Fractal image format (FIF), 128
Framegrabbers (image capturing systems), 111, 113, 120–123
Frames (images), 113, 114–116
Frequency converters, 212
Frequency response limitations of sensors, 103–104
Friction, 154, 167
FRL (filter-regulator-lubricators), 34, 35, 237, 244
Fuzzy logic controllers, 181, 187

Gain loss, 165
Gain of series of components, 153–154
Gas plasma displays, 24
Gate turn off (GTO) thyristors, 224

Gateways, 268
Gaussian filters, 131
Gear reducers, 40
Generators
    AC, 100–101, 102
    venturi vacuum, 31, 34
Ghosting of images, 131
Graphical analysis of servosystems, 158–161
Graphics displays, 25
Gray numbering system, 100
Grey scale reduction and thresholding, 124–127
Gripper flanges, 236
Grippers, 241
Group technology (GT) software, 270
GTO (gate turn off) thyristors, 224

Hall effect switches, 81
Hard automation, 7
Harmonic drive gear reducers, 40
Headers of messages, 214
Heat actuators, 21
Heaters, 21, 22
Hierarchical networks, 216
Histograms, 126, 127
Hydraulic actuators, 35–36, 238–239
Hydraulic power supplies, 36, 243–244
Hydraulic system safety considerations, 279
Hydraulic valves, 35–36
Hysteresis in control systems, 168
Hysteresis motors, 64

IC. *See* Integrated circuits (IC)
Image capturing systems (framegrabbers), 111, 113, 120–123
Image coding in vision systems, 128–129
Image enhancement, 130–132
Image tables
    input, 219
    output, 220
Images (frames), 113, 114–116
Impulse signals, 164, 165
Incremental optical encoders, 96–98
Index registers, 199
Indexer motors, 38, 39
Indexing tables, 37, 38–39
Induction motors, 52, 60–64, 237
Inductive proximity sensors, 82, 83

Inductive transducers, 92
Industrial process control overview, 6–8
Inertia, 154, 167
Ink jet printers, 25
Input image tables, 219
Input/output (I/O) devices
   analog, 208–212
   communications and protocol, 214–218
   introduced, 204–205
   modules for PLC, 220
   multiplexers, 121, 213
   parallel, 205–207
   in robots, 247–248
   serial, 213
Integral controllers, 175
Integrated circuits (IC)
   application specific integrated circuits (ASIC), 199
   temperature detectors, 86, 87
Integrated services digital network (ISDN), 215
Integrators, 223, 224
Intelligent peripherals, 194, 205
Interfacing
   controllers to controllers, 14–15
   controllers to other components, 15–16
   introduced, 12–14
   signal conditioning, 12–16, 204, 205
Interferometer type sensors, 95–96
Interleaved scanning, 120
Interlocks, 279
Internal memory, 198
International Standards Organization (ISO)
   Open Systems Interconnection (OSI) model, 14–15, 217–218
   series 9000 standards for statistical process control (SPC) quality control procedures, 278
Interpreted programs, 204
Interrupt-driven programming, 189
Interrupt mask bit, 200
Interrupts, 252
Inverse transform method, 131
I/O. *See* Input/output (I/O) devices
ISDN (integrated services digital network), 215
Islands of automation, 4, 264–265, 278
ISO. *See* International Standards Organization (ISO)
Iterative modification technique, 133, 135

Just in time (JIT) manufacturing, 276–278

Kanbans, 276, 277

Labor force and automation, 284–287
Ladder logic, 195, 219–220
LAN (local area networks), 268
Laplace transformation, 156–157
LCD (liquid crystal displays), 23–24
LED (light emitting diode) displays, 21–23
Left hand rule for motors, 50–51
Lenses in cameras, 116–117
Light actuators, 21–25
Light beam sensors, 84, 85, 86
Light emitting diode (LED) displays, 21–23
Light sensors, 84–85
Limit switches, 79–81
Line drivers, 206–207
Linear variable differential transformers (LVDT), 92–93
Linear vibratory feeders, 37, 39
Linearity of sensors, 78, 79
Liquid crystal cells, 23
Liquid crystal displays (LCD), 23–24
Local area networks (LAN), 268
Local counting method, 133, 136
Lubrication in compressed air delivery systems, 30–31
LVDT (linear variable differential transformers), 92–93

Magnetic flowmeters, 89
Magnetic sensors, 100, 101
Magnetostrictive position sensors, 93, 94
Magnets, permanent, 64
Mainframe computers, 199
Manufacturer's Automation Protocol (MAP), 15, 218
Manufacturing Resource Planning (MRP II) software, 196, 270
MAP (Manufacturer's Automation Protocol), 15, 218
Master production schedule (MPS), 270
MCU (microcontroller units), 199
Memory
   flash, 199
   internal, 198
   random access memory (RAM), 113, 129–130, 198, 245, 246–247
   read only memory (ROM), 198–199, 245–247
   in robots, 245–247
   in vision systems, 113, 129–130

Microcomputers, 199
Microcontroller units (MCU), 199
Microprocessor units (MPU), 199, 244–245
Microswitches, 80
Minicomputers, 199
Modems, 215
Momentary contact switches, 79
Monotonic instability, 162
Monotonic stability, 161–162
Motor controllers, 48–50
Motors
    AC. *See* AC motors
    air, 31
    armatures, 48, 50
    bearings, 48
    braking, 57, 68
    brushless DC, 70–73, 237
    capacitor run, 64
    capacitor start, 63–64
    commutators, 48
    compound wound, 55
    construction of, 48–50
    control of
        AC motors for control applications, 58–68
        DC motors for control applications, 54–58, 59
        electronically commutated motors, 68–73
        introduced, 53
    DC. *See* DC motors
    electronically commutated, 68–73
    history of, 47–48
    hysteresis, 64
    indexer, 38, 39
    induction, 52, 60–64, 237
    left hand rule for, 50–51
    moving coil, 54
    permanent capacitor, 64
    permanent magnet, 54, 237
    printed circuit, 54
    reactor start, 63
    reluctance, 64
    resistance start, split phase, 62, 63
    right hand rule for, 51, 52
    in robots, 236–237
    rotors, 48
    series wound, 54–55
    shaded pole, 62, 63
    shunt wound, 55
    silicon controlled rectifier (SCR), 70, 71
    slip rings, 48
    split phase, 62–64
    squirrel cage, 61
    stators, 48
    stepper, 50, 69, 236–237
    stopping of. *See* Stopping of motors
    synchronous AC, 50, 64
    theory of operation, 50–52
    timing, 69–70
    torque, 27
    types, 52–53
    universal, 64–65
Moving coil motors, 54
MPS (master production schedule), 270
MPU (microprocessor units), 199, 244–245
MRP II (Manufacturing Resource Planning) software, 196, 270
Multiple output control, 185
Multiplexers, 121, 213

Natural (resonant) frequency, 103, 160, 172
NC (normally closed) switches, 79
NC (numerical control) equipment, 41–42, 266
Near-robots, 232–233
Needle diagrams, 139, 142, 143
Nested control loops, 183
Network layer (OSI model), 218
Network service layers (OSI model), 218
Networks and communications, 214–218, 268
NO (normally open) switches, 79
Non-contact limit switches, 81
Non-contact presence sensors, 81–86
Normally closed (NC) switches, 79
Normally open (NO) switches, 79
NPN sensors, 82–83
NPN transistors, 222
Numbering systems
    binary, 99–100
    Gray, 100
Numerical control (NC) equipment, 41–42, 266
Nutating disc flowmeters, 89
Nyquist plots, 160, 161
Nyquist stability criterion, 161

Object recognition, 132–140
Offline programming, 248
Offset signals, 185

On/off control, 170, 171
Op amps, 57, 100, 101, 207, 222–224. *See also*
    Controller algorithms
Open architecture of computers, 13–14, 16
Open loop control systems, 6, 7, 151, 156
Open Systems Interconnection (OSI) model, 14–15,
    217–218
Operating systems (OS), 202–203, 246
Operational amplifiers, 57, 100, 101, 207, 222–224.
    See also Controller algorithms
Optical encoders, 96–100
Optical proximity sensors, 83–85
Optical switches, 206
Optical transducers, 92
Opto-isolators, 206, 207
OS (operating systems), 202–203, 246
Oscillating instability, 162, 165–166
Oscillating stability, 162
OSI (Open Systems Interconnection) model, 14–15,
    217–218
Output image tables, 220
Overshoot, 158, 172

Packets, 214
Panel cutters, 41
Parallel grippers, 241
Parallel I/O, 205–207
Path control of robots, 249–252
Pathnodes, 251–252
Pattern recognition, 132–140
Payback period, 280–283
PB (proportional band), 173
PC (personal computers), 13–14, 199
PCMCIA (Personal Computer Memory Card
    International Association) cards, 199
PD (proportional derivative) control, 175–177
PE (proportional) control, 170–173
Peripherals
    intelligent, 194, 205
    in robots, 240, 242–243
    in workcells, 193–194
Permanent capacitor motors, 64
Permanent magnet DC motors, 54, 237
Permanent magnets, 64
Personal Computer Memory Card International
    Association (PCMCIA) cards, 199
Personal computers (PC), 13–14, 199
Phase lag, 165

Phase locked loop (PLL) speed control systems, 101,
    102
Photodiodes, 91
Physical layer (OSI model), 218
PI (proportional integral) control, 173–175
Pick and place robots, 232–233, 238
PID (proportional integral derivative) control,
    178–181
Piezoelectric force generators, 25
Piezoelectric strain sensors, 88
Pinhole cameras, 116, 117
Pitch, 235
Pitot type flow meters, 89
Pixels, 110
PLC. *See* Programmable logic controllers (PLC)
PLL (phase locked loop) speed control systems, 101,
    102
Plugging, 57
Pneumatic actuators, 237–238
Pneumatic system safety considerations, 279
PNP sensors, 82–83
PNP transistors, 222
Point to point robots, 250–251
Polarizing filters, 23, 24
Poppet valves, 31, 32
Position sensors, 89–100, 101
Potentiometers, 92
Power supplies, 36, 243–244
Power transmission components, 39–40, 41
Predictive control systems, 184–185
Predictive controllers, 175–177
Preprocessors, 113, 123–129
Presentation layer (OSI model), 218
Pressure switch arrays, 91
Pressure transducers, 86–88, 89
Printed circuit motors, 54
Printers
    dot-matrix, 26
    ink jet, 25
Process control overview, 6–8
Processors in vision systems, 113, 130–142, 143
Production ordering kanbans, 276, 277
Program counter, 199
Programmable logic controllers (PLC)
    in automated workcells, 259–261
    development of, 12–13, 14
    introduced, 187
    overview, 195–196
    PC versus, 219–220

Programmable logic controllers (PLC), cont.
    in pick and place robots, 232–233
Programmable read only memory (PROM), 198–199
Programs and programming
    application programs, 203–204, 246
    artificial intelligence programming languages, 253
    communications programs, 217–218, 248
    compiled programs, 204, 252–253
    decision support software (DSS), 269–270
    group technology (GT) software, 270
    interpreted programs, 204
    interrupt-driven programming, 189
    Manufacturing Resource Planning (MRP II) software, 196, 270
    offline programming, 248
    operating systems (OS), 202–203, 243
    programming languages for robots, 252–253
    servocontrol programs, 245
    terminal emulation software, 248
    user programs, 204
PROM (programmable read only memory), 198–199
Proportional (PE) control, 170–173
Proportional band (PB), 173
Proportional derivative (PD) control, 175–177
Proportional integral (PI) control, 173–175
Proportional integral derivative (PID) control, 178–181
Proportional valves, 35, 36
Proprietary standards, 15
Protocols, 15, 216, 217–218
Proximity sensors, 81–86
Pulse width modulation (PWM), 58, 59, 65, 67, 208
Pumps
    hydraulic, 243–244
    vacuum, 34
PWM (pulse width modulation), 58, 59, 65, 67, 208

Quality circles (QC), 277
Quantization of analog signals, 122–123, 124

R-2R digital to analog converters, 208
Radio frequency modems, 215
RAM (random access memory), 113, 129–130, 198, 245, 246–247
Ramp analog to digital converters, 209
Ramp signals, 164

Random access memory (RAM), 113, 129–130, 198, 245, 246–247
Range of sensors, 76, 77
Raster scan, 25
Ratio controllers, 185
Reactor start motors, 63
Read only memory (ROM), 198–199, 245–247
Recognition units, 114, 142–144
Reflectance maps, 137
Reflected waveform sensors, 93–94, 95
Region growing method, 133, 136, 141
Registers in CPUs, 199–200
Reluctance motors, 64
Repeatability of sensors, 78
Repeating signals, 165
Reset controllers, 173–175
Reset time, 175
Reset vector, 201
Residual errors, 172
Resistance of control systems, 169
Resistance start, split phase motors, 62, 63
Resistive temperature detector (RTD) sensors, 86, 87
Resolution of sensors, 77–78
Resolvers, 96, 97
Resonant (natural) frequency, 103, 160, 172
Retentive addresses, 220
Retroreflective sensors, 84, 85, 86, 93–94
Right hand rule for motors, 51, 52
Ring networks, 217
Robots
    central processing unit (CPU), 244–245
    components
        arm, 240–241
        controller, 244–248
        end effectors, 241–242
        introduced, 239–240
        peripherals, 240, 242–243
        power supply, 243–244
    continuous path, 251–252
    controlled path, 251
    degrees of freedom, 233–236
    development of, 13, 15
    direct drive, 237
    geometries
        additional degrees of freedom, 235–236
        introduced, 233
        three major axes, 233–235
    input/output (I/O) devices, 247–248
    introduced, 231–232

Robots, cont.
- levels of sophistication
  - introduced, 253
  - path control, 249–252
  - programming languages, 252–253
- memory, 245–247
- and near-robots, 232–233
- pick and place, 232–233, 238
- point to point, 250–251
- power sources
  - electric motors, 236–237
  - hydraulic actuators, 238–239
  - introduced, 236–237
  - pneumatic actuators, 237–238
- selective compliance articulated robot actuator (SCARA), 233–235
- teach pendants for, 247–248

Roll, 235
ROM (read only memory), 198–199, 245–247
Rotary actuators, 30
Rotary indexers, 38, 39
Rotary resolvers, 96, 97
Rotary synchros, 96
Rotary vibratory (bowl) feeders, 38
Rotors in motors, 48
RS 232 standard, 14, 214
RS 422, 423, and 485 standards, 214
RTD (resistive temperature detector) sensors, 86, 87
Run length coding, 129

Safety considerations, 279–280
Safety interlocks, 279
Sampling of analog signals, 121–122
Saturation in control systems, 168
Saturation values, 154
Scan sequence of PLC, 219, 220
Scanners
- bar code readers, 104–105, 266
- ultrasound, 94–95

SCARA (selective compliance articulated robot actuator), 233–235
Scene analysis, 140–142, 143
SCR (silicon controlled rectifiers), 224, 225
SCR (silicon controlled rectifier) motors, 70, 71
Selective compliance articulated robot actuator (SCARA), 233–235

Sensors
- acceleration, 101–102, 103
- alternating signal, 102–104
- as automation components, 8, 10–11
- error in, 76
- flow transducers, 88–89, 90
- force transducers, 86–88, 89
- frequency response limitations of, 103–104
- introduced, 76
- light, 84–85
- linearity of, 78, 79
- magnetic, 100, 101
- non-contact presence, 81–86
- NPN, 82–83
- PNP, 82–83
- position, 89–100, 101
- pressure transducers, 86–88, 89
- proximity, 81–86
- quality of, 76–78, 79
- range of, 76, 77
- repeatability of, 78
- resolution of, 77–78
- span of, 76, 77
- special purpose, 104–105
- switches as, 79–81
- temperature transducers, 86, 87, 88, 89
- velocity, 100–101, 102

Serial I/O, 213
ISO Series 9000 standards for statistical process control (SPC) quality control procedures, 278
Series wound motors, 54–55
Servocontrol programs, 245
Servogrippers, 236
Servomotors, brushless DC, 70–73
Servosystems
- analysis of
  - by graphical methods, 158–161
  - by transfer function, 156–157
- closed loop control systems, 151–155, 156
- control configurations. *See* Controller configurations
- controller and the controlled system, 181–182
- digital controllers, 186–189
- direct drive, 72
- introduced, 151
- response of
  - controller algorithms and system response. *See* Controller algorithms

Servosystems, cont.
    input signals, disturbance types, and system response, 163–166
    physical characteristics and system response, 166–170
    stable versus unstable response, 161–162, 165–166
Servovalves, 35–36
Session layer (OSI model), 218
Setpoints, 151–152, 163
SFC (shop floor control), 270
Shaded pole motors, 62, 63
Shop floor control (SFC), 270
Shunt field control, 56–57
Shunt wound motors, 55
Signal conditioning, 12–16, 204, 205
Signal isolation, 205–206
Silicon-based controllers, 221–225
Silicon controlled rectifier (SCR) motors, 70, 71
Silicon controlled rectifiers (SCR), 224, 225
Simplification
    introduced, 4, 276
    just in time (JIT) manufacturing for, 276–278
    statistical process control (SPC) for, 278
Single-phase AC motors, 61–62
Single pole (SP) switches, 80
Single throw (ST) switches, 79
Single-turn indexers, 39
Sinusoidal signals, 164, 165
Six step method, 67, 68
Slip in induction motors, 61
Slip rings in motors, 48
Social effects of automation, 284–287
Soft automation, 7–8
Software. *See* Programs and programming
Solenoids, 25–27
Solid state switches, 206
SP (single pole) switches, 80
Span of sensors, 76, 77
SPC (statistical process control), 104, 278
Speakers, 25
Speed control for DC motors, 55–57
Spike models, 140
Split phase motors, 62–64
Spool valves, 32, 33
Spooled input, 213
Spring returns for valves, 32
Spring-type pressure sensors, 87, 88
Squirrel cage motors, 61
ST (single throw) switches, 79
Stable versus unstable response, 161–162, 165–166
Standards
    actuator and sensor connections to controllers, 15–16
    Automatic Identification Manufacturers Uniform Symbol description 2 (AIM USD-2), 105
    carrier sense multiple access with collision detection (CSMA/CD), 216
    Code 39, 105
    fiber digital data integration (FDDI), 216
    ISO series 9000, for statistical process control (SPC) quality control procedures, 278
    Open Systems Interconnection (OSI) model, 14–15, 217–218
    proprietary, 15
    RS 232, 14, 214
    RS 422, 423, and 485, 214
    Universal Product Code (UPC), 105
Star networks, 216
Static friction, 167
Statistical process control (SPC), 104, 278
Stators in motors, 48
Steady state errors, 158, 172
Step signals, 163–164
Stepper motors, 50, 69, 236–237
Stopping of motors
    AC, 68
    DC, 57
Strain gauges, 87–88
Structured lighting, 114, 115
Successive approximation analog to digital converters, 209, 210
Supercomputers, 199
Switch arrays, 91, 92
Switches
    contact bounce in, 80–81
    debouncing, 81
    detent, 79
    double pole (DP), 80
    double throw (DT), 79
    Hall effect, 81
    introduced, 10
    limit, 79–81
    microswitches, 80
    momentary contact, 79

Switches, cont.
   non-contact limit, 81
   normally closed (NC), 79
   normally open (NO), 79
   optical, 206
   as sensors, 79–81
   single pole (SP), 80
   single throw (ST), 79
   solid state, 206
   triple pole (TP), 80
   zero speed, 57
Synchronous AC motors, 50, 64
Synchronous communications, 214
Synchros, 96
Systems houses, 42

Teach mode for vision systems, 144
Teach pendants for robots, 247–248
Technical and Office Protocol (TOP), 15
Temperature transducers, 86, 87, 88, 89
Templates for vision systems, 132–133, 134, 135
Terminal emulation software, 248
Tessellated spheres, 139–140
Thermistor temperature sensors, 86, 87
Thermocouples, 86, 87
Thresholding in vision systems, 125–126
Through-beam sensors, 84, 85, 86
Thyristors, 224–225
Timing belts, 40
Timing motors, 69–70
Token passing bus networks, 217
Tone generators, 25
TOP (Technical and Office Protocol), 15
Torque motors, 27
Toyota just in time (JIT) manufacturing, 276–278
TP (triple pole) switches, 80
Tracking analog to digital converters, 210–212
Transducers
   flow, 88–89, 90
   force, 86–88, 89
   inductive, 92
   introduced, 10
   optical, 92
   pressure, 86–88, 89
   temperature, 86, 87, 88, 89
Transfer function
   of servosystems, 156–157
   of signal conditioner blocks, 154

Transformers, linear variable differential transformers (LVDT), 92–93
Transistors
   bipolar, 221–222
   field effect transistors (FET), 222, 223
Transport layer (OSI model), 218
Triacs, 225
Triple pole (TP) switches, 80
TTL level, 204
Type 1 and type 2 errors, 131, 143

Ultrasound scanners, 94–95
Uncoordinated point to point robots, 251
Unemployment and automation, 284–287
Unit error signals, 163–165
Universal motors, 64–65
Universal Product Code (UPC), 105
Unstable versus stable response, 161–162, 165–166
UPC (Universal Product Code), 105
User programs, 204

Vacuum cups, 241
Vacuum pumps, 34
Vacuum supplies, 34
Valves
   air-control valves and actuators, 31–33, 34
   detent, 32–33
   hydraulic, 35–36
   poppet, 31, 32
   proportional, 35, 36
   servovalves, 35–36
   spool, 32, 33
   spring returns for, 32
Vane type flow sensors, 89
Variable voltage inversion (VVI), 67, 68, 208
Velocity sensors, 100–101, 102
Venturi vacuum generators, 31, 34
Vibration sensors, 102
Vibratory feeders, 37–38, 39
Vidicon cameras, 119–120
Vision systems
   components
      camera, 113, 116–120
      image (frame), 113, 114–116
      image capturing system (framegrabber), 111, 113, 120–123
      introduced, 113–114
      memory, 113, 129–130
      output interface, 112, 114, 144
      preprocessor, 113, 123–129

Vision systems, cont.
    processor, 113, 130–142, 143
    recognition unit, 114, 142–144
  introduced, 110
  simple, 110–112
Vortex shedding flow sensors, 89
VVI (variable voltage inversion), 67, 68, 208

WAN (wide area networks), 268
Water-jet cutters, 41
Wide area networks (WAN), 268
Windowing in vision systems, 127–128
Withdrawal kanbans, 276, 277
Workcell controllers, 194–196, 258–264, 265
Workcells
  automated, 256–264, 265
  just in time (JIT) manufacturing and, 276–278
  peripherals in, 193–194
Workforce and automation, 284–287
Wound field AC induction motors, 61
Wound field DC motors, 54–55

X-Y tables, 41, 42, 43

Yaw, 235

Zener diodes, 221, 222
Zero speed switches, 57